PRACTICAL GUIDE

AGRICULTURAL VALUATIONS
A
PRACTICAL GUIDE

(Second Edition)

by

R. G. WILLIAMS
F.R.I.C.S. F.A.A.V.

1991

A member of Reed Business Publishing Group

THE ESTATES GAZETTE LIMITED
151 WARDOUR STREET, LONDON W1V 4BN

First Published 1985
Second Edition 1991

ISBN 0 7282 0161 5

Typeset by BP Integraphics Ltd, Bath
Printed in Great Britain at The Bath Press, Avon

Foreword

When Gwyn Williams told me he was writing a revision to his book 'Agricultural Valuations—A Practical Guide', that is exactly what I expected from the author, who has completed 36 years of very successful practice in agricultural valuations.

When I was sent the proof it was, therefore, no surprise to me that here we have a really helpful and very practical guide. The book contains a wealth of well-considered, practical advice; not only for new entrants into the Profession but also for experienced valuers.

The Central Association of Agricultural Valuers' examination system is a test of professional competence after a candidate has gained practical experience in the field of agricultural valuations. When the book was first published in 1985, it was well received by a wide spectrum of readers and, in my opinion, the new book is very much more than just a revision. There are completely new chapters dealing with such matters as management agreements and milk quotas which are becoming more and more important following the change of emphasis in agriculture from maximum food production to the use of the countryside including food production.

Gwyn Williams is an acknowledged expert in the field of agricultural valuations on the Welsh borders. When I was asked to read the book coming from the corn belt of the east, I wondered how I would react. I am sure my colleagues in the east will find the book just as helpful as those from the west, in spite of there being various matters where emphasis would differ widely between the livestock and arable areas.

In the review to the first edition, Ted Potter stressed the importance of the book for newcomers into the Profession. I personally doubt whether many older practitioners could read the book without gleaning much useful information. I would stress to younger readers that I am certain they will find great benefit from becoming Members of the Central Association of Agricultural Valuers, to be kept up-to-date with practical professional problems that are continually arising in the fast-changing agricultural scene. The

Committees of the Central Association comprise people like Gwyn Williams from all parts of England and Wales, covering a very wide field of expertise in the affairs of the countryside. It is a strength of the Association that these people are prepared to pass on to others the benefit of their experience in both the Association's newsletter and in their numbered publications. I am sure that Gwyn Williams would be the first to acknowledge the use of information and figures which have come from the C.A.A.V. I think we all, in practice, owe a great debt of gratitude to Gwyn Williams for the time he has obviously taken in preparing his book and for the way he has written the book in such a readable fashion.

C. R. WRIGHT, B.Sc. (Est.Man) London, F.R.I.C.S., F.A.A.V.
President, C.A.A.V. 1989/90
Fenn Wright Spurlings,
Colchester, Essex.

Preface to the First Edition

This book started its life as a few articles dealing with specific types of valuation, written for the benefit of a number of young agricultural valuers, to assist in their professional examinations. I have been encouraged by several practitioners to develop these articles into book form and this publication is the result. Many young agricultural valuers have, over the last few years, had considerable difficulty in obtaining proper experience in agricultural valuations. The problems have increased as firms have become specialised, the work-load of principals has changed and increased, resulting in little time being available to instruct young valuers. I was further encouraged to publish this book as there is no similar, up to date publication, currently available.

Some valuers may consider that some of the figures given in the example claims are high and in some cases items may even appear to be duplicated. Nevertheless, it is the duty of any claimant's valuer to prepare a proper and full claim as he may otherwise be negligent, if he does not do so. Valuation is not an exact science and virtually every claim made by a practitioner is subsequently negotiated.

This book is essentially a practical guide drawn from more years experience in the profession than I care to remember. The views expressed and methods of valuation are, as I use, and interpret matters. Whilst every effort has been made to ensure the accuracy of the information given and opinions expressed, I cannot, however, accept any responsibility for any inaccuracy of these interpretations and valuation methods. I hope this book will prove to be of value as a practice guide to young and older valuers alike.

I would like to take this opportunity of expressing my appreciation to the President and Council of the Central Association of Agricultural Valuers for allowing me to use some of their published costings and also their encouragement in writing this book.

My special thanks go to Mr. Alan N. Lane, A.R.I.C.S., C.A.A.V. of Knutsford who has corrected the manuscript, written the article on Pigs and also given me many valuable suggestions. Also to the following who have given valuable advice:

Mr. Robert Thomas, M.A., F.R.I.C.S., C.A.A.V. Cowbridge; Mr. Alan Brown, F.R.I.C.S., C.A.A.V. Newport; Mr. Malcolm Williams, C.D.A. Ross-on-Wye and a leading compensation valuer.

I further wish to thank the following members of my own firm for checking the calculations, advice and making suggestions:

Mr. Nigel Morris, A.R.I.C.S., N.D.A., C.A.A.V.; Mr. Michael Taylor, B.Sc, A.R.I.C.S. and my son Mr. Richard Williams, B.Sc A.R.I.C.S. A.S.V.A. C.A.A.V.

Finally thanks go to four very patient Secretaries who have typed and re-typed this work.

October 1984 *R. G. Williams*
COLES, KNAPP & KENNEDY
Ross-on-Wye.

Preface to the Second Edition

The first edition was intended to be for the use of students of land management and agricultural valuation. I was pleasantly surprised at the welcome given to the book, not only by students but also by established practitioners, colleges, bank managers etc. The Central Association of Agricultural Valuers also honoured me in 1985 by awarding me the 'Kenneth Glenny' Prize for writing this book. The book obviously answered a need.

New law and consolidating law has necessitated the provision of this second edition which is fully updated and extended with a new chapter on arbitrations. A chapter on Management Agreements has been written with the help of the N.C.C. David Carter has written a chapter on Milk Quotas and Paul Wright a chapter on Soils. I am most grateful to them all for their kind help and interest.

I would like to thank my secretary Miss C. Morgan for her secretarial help, Mr. A. J. Lyke, A.S.V.A. F.A.A.V. for updating the Glossary of general information, Mr. Malcolm Williams, C.D.A. of M.S.F. Ltd for advice on sprays etc, a noted compensation valuer (who again prefers to remain nameless!) for checking the chapter on compensation and compulsory purchase. Also Mr. A. H. Durrant for reading and amending the chapter on Arbitration.

Finally, a word of thanks to the President and Council of the Central Association of Agricultural Valuers for allowing me to use some of their costings and other information, and also a special thank you to Mr. C. R. Wright B.Sc F.R.I.C.S. F.A.A.V., their President for writing the Foreword.

R. G. Williams
Coles, Knapp & Kennedy
Ross-on-Wye

April 1990

Contents

CHAPTER ONE

History

1.1 Agricultural Valuations, as commonly known, can be said to have started as the assessment of tenant right payments, when a tenant quitted an agricultural holding. Such a valuation was known as a tenant right valuation. Tenant right compensation started around 1775 when there was considerable agricultural activity and progress in the country.

1.2 This compensation was paid for on a basis of custom, which developed over the whole country and often varied from county to county and in some cases from estate to estate. The work of assessing the claim was done by farmer valuers, who were well known and respected farmers in the areas in which they lived. These men eventually formed county and area Tenant Right Valuers Associations. Most of these were formed after the passing of the Agricultural Holdings Act, 1875, when it became necessary to agree to a uniform basis, in their district, of assessing compensation for the various matters of claim.

1.3 The oldest known Valuers' Association was founded in the County of Suffolk in 1847. In 1910 a number of these Valuers' Associations joined together to form the Central Association of Agricultural Valuers, which, today, is the foremost agricultural and tenant right valuers' professional organisation in the land.

1.4 The items to be paid for as tenant right, under custom, were variously: hay, straw, dung, growing crops, acts of husbandry (tillage operations), compensation for feedingstuffs consumed by animals, fallows, offgoing or awaygoing crops, pre-entry (i.e. entry on to the holding in advance of the date of the tenancy, e.g. to plant crops), boosey pasture and holdover (i.e. right to keep a small field for a cow or horses from quitting day of say Candlemas (2nd February) until Whitsunday (15th May)).

1.5 With the passing of the Agricultural Holdings Act, 1923 and particularly the Agricultural Holdings Act, 1948, with their intricacies, the farmer valuers began to die away and the work was undertaken by professional agricultural valuers. However, farmer valuers did not completely disappear until about 25 years ago.

1.6 Prior to this, assessments of the value of land were made for the relief of the poor following the passing of the Poor Relief Act of 1601.

1.7 The work of the professionals was further increased during the railway construction era of the 19th Century when considerable areas of land were compulsorily acquired and compensation had to be assessed under the provisions of The Land Clauses Consolidation Act 1845.

1.8 The introduction of Death Duties by Mr David Lloyd George in his budget of 1910 further increased the work of an agricultural valuer as the landed gentry were assessed also on the value of their estates.

1.9 Nevertheless, most of the work of an agricultural valuer, as we know it today has been extended considerably since the end of the second world war and is very specialist in nature. Nowadays, it covers not only tenant right valuations but also, rental valuations, agricultural property valuations for many purposes including compulsory acquisition, purchase, sale, Inheritance Tax, Capital Gains Tax. Also compensation claims for easements (gas, water and oil pipelines, electricity lines), compensation for opencast and other mineral extraction, milk quotas, SSI's, etc..

CHAPTER TWO
Definitions etc

2.1 This book deals very substantially in termination of tenancy claims and it is expected that the practitioner has a working knowledge of the Agricultural Holdings Act 1986. Unless otherwise stated, the valuation examples given in Chapters 3–11, are all on the basis prescribed by the 1986 Act and, in particular, S.I. No. 809 The Agriculture (Calculation of Value for Compensation) Regulations 1978. S.I. No. 822 The Agriculture (Calculation of Value for Compensation) (Amendment) Regulations 1981 and the latest amending S.I. No. 1475 The Agriculture (Calculation of Value for Compensation) (Amendment) Regulations 1983.

2.2 In computing the relevant valuations, it is well to be reminded of the following which are of importance:

2.2(i) Pre 1st March 1948 Tenancies

Some of these still exist. Compensation for tenant right i.e. growing crops, harvested crops, root crops, seeds sown, pasture, sod fertility, acclimatisation of hill sheep etc., will not be paid under the provisions of Part II of the Eighth Schedule of the 1986 Act, unless before the termination of the tenancy, the tenant serves a written Notice under Para. 6 of Schedule 12 electing that those provisions shall apply to him. If no such Notice is served, compensation for tenant right will be payable under the custom of the district and the provisions of the tenancy agreement.

2.2(ii) Consuming Value

This is defined in S.I. 1978 No. 809 Part II 8(3) as being the market value for consumption on the holding by agricultural livestock of hay, fodder, crops, straw, roots and other crops or produce of good quality less the manurial value, on a 'no crop off' basis as prescribed by Tables 5(a) & (b) of S.I. 1983 No. 1475.

Local Valuers' Associations normally fix consuming values for various produce and crops twice a year. These figures reflect the deduction for the manurial values of the crops. The values fixed normally refer to the best quality, situated in a good position etc.

2.2(iii) Market Value

There is no official definition of market value but this is considered to be what the open market would give for any commodity, without restriction on where it was consumed e.g. the market value of a bay of hay is what is considered could be obtained for the same if sold by open auction. Likewise, the market value of an area of swedes is what the open market would give for the same, either for folding off or on harvesting and removal from the holding.

2.2(iv) Spring Tenancy (S.I. 1978 No. 809. Part II 8/2)

This means a yearly tenancy which commenced between 1st January and 30th June, inclusive.

2.2(v) Autumn Tenancy (S.I. 1978 No. 809. Part II 8/2)

This means a yearly tenancy which commenced between 1st September and 31st December inclusive.

2.2(vi) Enhancement Value (S.I. 1978 No. 809 Part II 8/1)(c)

This applies to *autumn sown crops* where the land is held under a *spring tenancy* or *grass and clover seeds* which are sown on the land held under a *spring or autumn tenancy* where no crop has been taken at the termination of the tenancy.

The enhancement value is an additional amount payable which represents the enhancement of the value of that crop to an incoming tenant but this shall not exceed the rental value at the termination of the tenancy.

2.3 TENANT RIGHT VALUATIONS

It is stressed that in all the Chapters dealing with tenant right there is no contract between the outgoing tenant and incoming tenant. The only contract the outgoing tenant has is with the landlord and a tenant right valuation is in theory carried out between the outgoing tenant and the landlord. Normally, however, one valuer acts for the outgoing tenant and another valuer acts for both the landlord and the incoming tenant (if there is one).

2.4 CLAIMS

Under Section 83(2) of the Agricultural Holdings Act 1986 Landlords and Tenants Claims must be notified before the expiration of two months (i.e. after) the termination of the tenancy. Under Section 83(4) of the 1986 Act eight months are allowed for settlement, failing which, the matter shall be determined by Arbitration under the provisions of Section 84 and Schedule 11 of the Act. Note, however, that if the landlord is claiming for general deterio ration of the holding, he must give the tenant notice of his intent to claim under S.72 of the Act no later than one month *before* the termination of the tenancy.

CHAPTER THREE

Produce

3.1 PRODUCE, comprises harvested hay, straw, silage and roots.

3.2 HAY

Distinguish between seeds and meadow hay. Good meadow hay is fine hay from old pastures and is very suitable for calves and sheep. Seeds hay is from Ryegrass, Timothy & Clover mixtures. Best hay should be leafy, bright, green, smell sweet, be cut young and not be fibrous. Hay suitable for horses has all of these qualities but is seeds hay and free from dust. Avoid fousty, dusty, fibrous, rain washed and overheated hay. Baling too soon and too tightly can spoil hay.

3.2(i) Valuation

3.2(ii) Standard Bales

Weigh say six bales in a bay, preferably from different layers and get average weight in kgs. Count number of bales, multiply weight by number of bales and divide by 1000 = number of tonnes, e.g. 920 bales av. 20.4 kg in bay

$$\frac{920 \times 20.4}{1{,}000} = 18.768 \text{ tonnes}$$

Volume Method

Sometimes it is not possible to count bales. An alternative method is to obtain the cubic content of the stack in M^3 and divide by an appropriate density figure:-

Very compact:	6–7 M^3 per tonne
Medium:	8–9 M^3 per tonne
Unsettled:	10–11 M^3 per tonne

Practice Checks

Small bales of hay often approximate at 50 bales per tonne. Medium and heavy bales often weigh from 18 to 22 kg per bale. Discount any weathered bales on exposed side of barns and also possibly half or sometimes the whole of the bottom layer as this might be damp. Look out for voids in the stack. The bottom layer has usually more bales than the other layers since they are laid on edge. As a rule a bay 24′ (7.3 m) × 15′ (4.5 m) × 18′ (5.4 m) will contain approximately 20 tonnes.

3.2(iii) Big Bales

In recent years big balers have become very popular especially on the medium and larger farms. Two types are generally used:

(a) Round Balers

These are usually of two types each producing a cylindrical bale of about 1.5 m wide × up to 1.8 m dia. or 1.2 m wide × up to 1.5 m wide. Note the density of bale made can be controlled and sometimes as a result bales may be very dense and thus heavier. Usually one round bale contains between 8 and 9 small bales.

(b) Rectangular Balers

Two types are in use viz a low density baler producing a bale about 1.5 × 1.5 × 2.4 m in size and a high density baler producing a bale of about 1.2 × 1.3 × 2.4 m. Often the weight of hay in big rectangular bales made by a high density baler is anything between 600 & 700 kg.

3.2(iv) Valuing big bales presents a major problem to an agricultural valuer since sample bales cannot easily be weighed. However, it is essential that say three or four bales are weighed and an average weight obtained. Often this means that they have to be transported to the nearest available weighing facility.

When the weight has been determined, the value can be assessed.

This may be on a consuming basis (the Local Agricultural Valuers Association may fix the values) which may need adjusting having regard to:-

(*a*) the quality of the hay.
(*b*) the quantity does not exceed the quantity reasonably required for the system of farming provided.
(*c*) how convenient it is for use.
(*d*) how well it is stacked and protected.

3.2(v) Points to Note

Big bales have in recent years become very popular and are regularly used. Nevertheless:

(*a*) Big bales are not as easy to handle as small bales. They must be handled mechanically.
(*b*) Not every farmer has the desire to use big bales and some prefer small ones.
(*c*) They may have been left out in the fields longer than small bales. The quality of the hay suffers in wet weather conditions.
(*d*) If the bales are out in the open but covered with polythene sheeting, unless placed on concrete they are likely to be damp and there could be considerable wastage.

3.3 STRAW

Wheat straw is usually used for bedding only. Spring barley straw is best for feeding. Winter barley straw is of secondary feeding value as it is brittle. Oat straw is also used for feeding but winter oat straw also tends to be brittle. The most valuable straw is, therefore, spring barley straw.

Feeding straw should be bright, not dusty, damp or weathered and the bales light. Wheat and oat straw bales are usually heavier than barley bales. Frequently there are about 60–80 small bales of barley straw to the tonne.

3.3 Valuation

3.3.2 Standard Bales

Method is as for hay subject to the following amendments on the density system (where bales cannot be counted).

Heavy bales:	10–11 M^3 per tonne
Medium:	12–13 M^3 per tonne
Light:	14–16 M^3 per tonne

3.3.3 Big Bales

Many of the comments made about big hay bales apply to straw and sample bales must be weighed. Often one round bale contains between 8 and 9 small bales. Often big straw bales are left out in the open but do not suffer, as a rule, if they are wrapped. However, there is considerable wastage and deterioration in quality after a few weeks exposure to rain and the weather generally where bales are not wrapped.

3.3.4 Valuation is on the same basis as for hay, using the relevant figures prescribed by the local Agricultural Valuers Association.

3.4 SILAGE

3.4.1 Clamp Silage

This should be valued on the basis recommended by the C.A.A.V. [Numbered Publication 154 as amended] and the Ministry of Agriculture. It should always be analysed and such analysis should be used in assessing both the density and the value of silage and research by A.D.A.S. has shown that there is direct relationship between dry matter percentage and dry matter density of clamp silage. The bulk density is obtained by the following formula:

$$\text{Kg per M}^3 = \frac{6500}{\text{DM\%}} + 400$$

$$\text{Lbs per cu ft} = \frac{410}{\text{DM\%}} + 25$$

D.M. = Dry Matter

Grass silage has a light green, greenish yellow or greenish brown colour and should smell pleasant, fruity or faintly acid and the texture should be firm. Dark brown or black silage is overheated. Olive green or dark green silage is underheated and has a strong rancid, musty smell. Sometimes secondary fermentation occurs and the silage becomes a brown colour and smells fishy or musty and is rather slimy in texture.

Care should be taken in measuring silage. If the bunker walls are lined with plastic sheeting at the side there is usually a good deal less waste than when unlined. Proper allowance should be made for wastage at the sides, the wedge and also the top where sometimes as much as 150 mm (6″) is waste.

Great care should be taken in analysing silage as often three or four cuts are taken and there can be considerable variation in the analysis of the various cuts. Often four samples are taken from four different points on the clamp and all are put in the same bag and analysed as one.

Precision chopped silage in clamps is usually easier to self-feed than single or double chop. Bagged silage is always cut by a mower (not precision chopped) and usually baled direct and then bagged.

3.4.2 Big Bale Silage

This should have sample analysis of at least 5% of the bales. In the case of bagged silage, where the analysis shows D.M. of less than 30% satisfactory fermentation may not be possible. The volume is calculated of each bag (cylindrical) by the formula $3.142 \times \text{radius}^2 \times \text{height}$.

If the plastic bags are punctured or badly sealed the silage is of little value and often the whole bag is ruined. Wrapped silage stores much better than bagged silage. The Outgoer is responsible for weighing a sufficient number of bales to establish the weight.

3.4.3 Maize Silage

This is an expensive crop to grow and often is harvested in wet conditions in mid to late autumn. It occupies the land for a long period (late April to October) and this land cannot often be used for any other purpose (contrast the use of land for grass silage and subsequently for grazing or for ploughing and planting roots or winter corn after grass silage crops are taken) that particular year. Despite this it can yield well and is a good break crop.

It can yield from 15 to 25 tonnes per acre. It has high digestability ('D' value of 60 to 70) but is low in protein.

3.4.4 Valuation

The C.A.A.V. recommend that silage should be valued by reference to the price per tonne fixed by Local Agricultural Valuers Associations for silage with a 10.5% M.E. (Metabolisable Energy) and 23% Dry Matter. Once this figure is fixed the price of all silage can be calculated using the following formula:-

Value of sample silage = Value fixed for silage of analysis as above

$$\times \frac{\text{\% M.E. of sample}}{10.5} \times \frac{\text{\% Dry Matter of sample}}{23}$$

Example (a)

A clamp measuring 25 m × 13 m × 2 m has an analysis of 12% C.P., 26% Dry Matter and 11.2% M.E. The local Valuers Association have a fixed price of £18 per tonne for 10.5% M.E. and 23% D.M. Silage value per tonne is therefore:

$$£18 \times \frac{11.2}{10.5} \times \frac{26}{23} = £21.70 \text{ per tonne}$$

Some silages are low in protein and it is recommended that those with crude protein below 14% in the silage dry matter should be reduced in value on the basis of for each 1% decrement of crude protein (C.P.) in the silage D.M. the reduction should be calculated using the formula:-

$$(\text{Soya price—Barley price}) \times \frac{\text{D.M. \% of Silage}}{4000}$$

Example (b)

$$\text{Reduction} = (£165 - £107) \times \frac{26}{4000} = £0.37$$

$$\text{Value calculated} = £21.70 - 2 \times £0.37 = £20.96 \text{ per tonne}$$

Example (c)

Value of clamp analysis and size in (a) above is:

$$\text{kg per M}^3 = \frac{6500}{26} + 400 = 650$$

Valuation is therefore 25 m × 13 m × 2 m = 650 m^3

$$650\,m^3 \text{ @ } 650\,kg = 422.5 \text{ tonnes @ £20.96}$$
$$= \text{£8,855.60 say £8,850}$$

Occasionally silage fermentation has not proved satisfactory and the recommended formula may over value the silage. The valuer should check on the silage fermentation by visual inspection and by reference to the silage analysis which should comment if the fermentation is unsatisfactory and in such cases an appropriate deduction should be made. Unsatisfactory fermentation will affect the preservation and feeding quality.

3.4.5 Tower Silage or Haylage

In practice a tower silo of 20′ diameter has a capacity of 2 tonnes *dry matter* per foot of depth of silage, a 24′ diameter silo has a capacity of 2.75 tonnes per foot and a 30′ diameter silo about 4.4 tonnes dry matter.

It is strongly recommended that a valuer should study the silo capacity given by the manufacturer.

A.D.A.S. advisers recommend that the *dry matter* capacity (note this is different from the actual capacity—dry matter content of fresh silage is taken at 40%) can be calculated on the following formula:

$$\text{D.M. capacity (tonnes)} = \frac{6.08 \times d \times 2h(17.4 + h)}{1{,}000}$$

d = diameter in metres
h = settled silage height in metres

Often it works out at 1.55 to 1.77 M^3 per tonne (560 to 650 kg per M^3) at 45 to 50% dry matter. (240 kg dry matter per M^3)

In computing the valuation, it is necessary to have the haylage analysed. Difficulty can arise in getting samples from a tower silo except at the bottom and before a valuation is settled, delay may arise in having to wait before further samples can be taken for analysis. Having obtained suitable analysis, it is suggested that the valuation should be done using the same formula as for clamp silage. Normally haylage has a much higher dry matter content than clamp silage and should have a much higher feeding value and thus, a higher value per tonne than silage.

3.4.6 Practice Note

No manurial value is given for silage or haylage in the regulations. This is because Table 5(a) of S.I. 1983 No. 1475 (as all previous regulations) deal with feeding stuffs consumed on the holding but not produced thereon (except corn). Thus the tables only deal with feeding stuffs that are bought and sold. It is expected that in future tables a consuming value for silage will be given. For the time being, where possible and on all sale valuation claims it should be stipulated that silage and haylage should be valued in accordance with C.A.A.V. N.P. 154 as amended.

3.5 HARVESTED ROOTS

The volume is calculated and the following are the amounts in M^3 and cu. ft. per tonne.

	M^3 per tonne	*Cu. Ft. per tonne*
Mangolds	1.78	64
Swedes	1.78	64
Potatoes	1.55	55
Fodder Beet	1.78	64

Local Valuers Associations, at their price fixing meetings, sometimes fix prices per tonne for these roots (in areas where they are grown) and this is done on the basis of the market value less the manurial value in a 'no crop off' basis in accordance with the S.I.

CHAPTER FOUR

Growing Crops

4.1 These are valued on the basis of cost of seeds, fertilisers, cultivations, fallows and acts of husbandry plus, in the case of autumn sown crops on a holding held on a spring tenancy, or grass and clover seeds sown in a spring or autumn tenancy where no crop has been taken, an additional amount representing the enhancement value of the incoming tenant of the growing crop not exceeding the annual rental value of the area carrying the crop. Often, the rent paid for the appropriate period of the growth of the crop is added as the enhancement value.

4.2 It is necessary to know the cultivations that are normally carried out. These are usually much less costly on light land than on heavy clay land. Note the action of frost on heavy land ploughed over winter improves the tilth considerably. Also note the increasing use of minimal cultivation or no-ploughing techniques. (See below). Full details of all cultivations carried out should be obtained from the farmer. A specimen cultivation procedure on light land for, say, a winter barley crop in Herefordshire, may be:

Winter:
1 Plough
1 Power Harrow
1 Combine Drill (grain & fertiliser—low nitrogen compound)
1 Seed Harrow
1 Pre-emergent spraying for weeds (herbicide)

Spring:
1–3 Top Dressings
1 Spray for mildew
1 Spray for weeds (herbicide) may be done in one operation (if not done in the autumn)
1 Rolling (possible)

Alternatively, for a crop of winter wheat on heavy land:

Winter:
1 Plough
1 Roll

2 Disc Harrows
1 Harrow
1 Combine Drill (grain & fertiliser—low nitrogen compound)
1 Seed Harrow
1 Pre-emergent spraying for weeds (herbicide)

Spring: 1 Roll and possibly
1 Harrow
1–3 Top Dressings

Additionally, the following spraying may be done (Wheat):
1 spraying for weeds (herbicide) if not done in the autumn
1 spraying for rust
1 spraying for aphids
1 spraying for growth regulator

Note

Four sprayings are shown above; however, apart from weed spraying, the others may not be necessary.

4.3 NO PLOUGHING OR MINIMAL CULTIVATIONS

Some farmers, particularly on heavy soils, do not plough, and have found in practice that in using minimal cultivations the top layer of soil has an improved texture and workability and resultant yields are improved. An example of the technique used by an arable farmer on heavy clay in Worcestershire (all winter corn) is:

1 spray (if the land is foul) to kill weeds and, if pasture, the grass)
1 subsoiling (every three years)
2 or possibly 3 heavy disc harrows
1 Power Harrow (Roterra)
1 Combine Drill (cultivator type)
1 Harrow
1 Spray (either pre-emergent or post-emergent) (preferably the latter.

4.4 VALUATION

4.4.1 Winter wheat

The valuation is calculated in accordance with the costings issued by the C.A.A.V. An example of a winter wheat crop is as follows:

	£	£
Cultivations etc.	*per acre*	*per Ha*
1 plough	11.40	28.15
1 roll (single)	4.20	10.35
1 power harrow	7.30	18.05
1 combine drill	6.65	16.40
1 seed harrow	1.75	4.30
1 spray	2.60	6.40
Seeds (Costing £215 per tonne)		
185 kg		39.77
75 kg	16.12	
Fertilisers (9:24:24 @ £145 per tonne)		
308 kg		44.66
125 kg	18.12	
Spray, pre-emergent herbicide	10.52	26.00
Enhancement value	16.50	40.75
	95.16	234.83

4.4.2 Roots (in the ground)

Growing root and green crops (e.g. kale, stubble turnips) of a kind normally grown on a holding held under an *autumn tenancy* are valued on the basis of the average market value on the holding of good quality crops, less their manurial values as provided by S.I. tables. In practice, the prices are fixed by the local agricultural valuers' association in the autumn. Example valuations are (as at autumn 1989).

Marrow stem kale	£150 per acre	£370 per Ha
Kale (other)	£115 per acre	£284 per Ha
Stubble turnips	£75 per acre	£185 per Ha

The yield of swedes, mangolds and turnips in the ground is estimated, (this can be checked by cutting and weighing specimen areas), and all are valued in the autumn for 1989 @ £15 per tonne.

The above values are for best quality crops and are net of the deduction for manurial values.

4.5 GROWING CROPS—MARKET VALUE BASIS

4.5.1 Occasionally, crops are required to be paid for on crop value or market value basis i.e. what that crop will realise if sold or taken over, as a growing crop on the relevant date. Note, this is most likely to arise under the contracts of sale of farms or form part of agreed surrender terms.

In such valuations it is usual to assess the likely yield, price this out at market value and then deduct the cost of harvesting and also a percentage of risk (usually, dependent on stage of maturity of the crop, 10%–15%) in case it may never be harvested because of weather conditions or losses on lodging (crop being laid) Examples:

4.5.2. Growing crop of Barley

		per acre		*per Ha*
Grain yield 2½ tonnes @ £105		£262		£647
Straw yield 1 tonne @ £15		£15		£37
		£277		£684
less: Harvesting, Carting & Drying	£40		£98	
risk: 10% of gross	£28	£68	£69	£167
Crop Value		£209		£517

4.5.3. Growing crop of Swedes

	per acre	*per Ha*
Estimated net yield 14 tonnes @ £15	£210	£518
less: Harvesting—nil (to be folded off)	—	—
risk: nil (if healthy—but if suffering from disease say 25–50% + can be deducted	—	—
Crop Value	£210	£518

Here it is observed that although these values are prescribed by local valuer's associations, it my prove in practice very difficult to get these figures on the open market. Also where folded off, about 1/4 of every root is left in the ground as it cannot be eaten.

4.5.4 Growing crop of Marrow Stem Kale

Note that this crop is very susceptible to frost damage, and it is most unwise to agree to pay anything much for such crop after Christmas as often frosts render the crop valueless. Canston/Maris

Kestrel/Hybrid Kales are generally frost resistant and, as weather durable crops, are of far greater value.

	per acre	*per Ha*
Value of good crops (as fixed by local Agricultural Valuer's Association	£150	£370
less risk (if large area still growing in ground, say, on 1st December(—50%	£ 75	£185
Less—Harvesting	NIL	—
VALUE	£ 75	£185

CHAPTER FIVE

Tenant's Pastures

The basis of valuation for Tenant's pasture is either of two bases.

5.1 COST BASIS

This basis is to be used where no crop has been removed by mowing or grazing and when sown on land held under a Spring or Autumn tenancy. The value is the reasonable cost of seeds sown, cultivations, fallows and acts of husbandry performed, also taking into account any expenditure incurred solely for the benefit of the pasture before the removal of any crop in or with which this pasture was sown. Also enhancement value shall be added.

It is very costly to establish leys, as shown in this example.

		per Acre		*per Ha*
Seeds (4/5 year ley) say		£20.00		£ 49.40
Cultivations viz:				
1 Plough	11.40		28.15	
1 Power Harrow	7.30		18.05	
1 Drill	3.35		8.25	
1 Fertiliser Spinner	2.35		5.80	
2 Roll (gang)	4.60	£29.00	11.40	£71.65
		£49.00		£121.05
Note: Further harrowing or rolling may be necessary				
Fertilizers 100 Kg 20:10:10		£13.00		
247 Kg 20:10:10				£32.11
Enhancement Value say		£20.00		£49.40
Total Cost		£82.00		£202.56

Short Term Leys (i.e. of 1/3 years)—mainly of Ryegrass and possibly other grasses, cost almost as much to establish, the only exception being that the seeds sown, cost an average of £5–£6 per acre, (£12.50–£15 per Hectare) less.

If the seeds are *UNDERSOWN* (i.e. drilled in the Spring in, say, a growing Spring Barley crop), the labour costs will be much less.

	per Acre	*per Ha*
Seeds say	£20.00	£49.40
Broadcasting Seeds (Spinner)	2.20	5.43
Rolling	3.50	8.64
	£25.70	£63.47

However, the practice of undersowing Corn with young seeds is now less popular, many Farmers preferring to sow them direct, in the autumn, after cereals.

5.2. FACE VALUE BASIS

This basis is applicable where one or more crop has been taken by mowing or grazing. In assessing the value of any particular pasture, the following shall be taken into account:

(*a*) present condition. It is a good quality ley, well established, of a good seeds mixture, clean and free from weeds, especially grass weeds such as couch?

(*b*) has it been properly managed since sowing? Leys can be ruined by undergrazing.

(*c*) is there water available for stock?

(*d*) is the fencing good and stockproof?

(*e*) is it convenient of access and also in relation to the remainder of the farm?

(*f*) is it near the end of its useful life and how much longer will it last before it becomes unproductive?

(*g*) is the ley dominated by certain grasses and clovers? Cocksfoot is generally a coarse (though productive and drought resistant) grass and if not properly grazed, will dominate a sward. Clover adds considerably to the fertility of the land by nitrogen fixation. However, where there is an excess in a sward, it causes blowing of animals (bloat) which occasionally results in death.

If leys are properly managed, they can improve in quality after they have been established a few years, provided of course, the seeds mixture is satisfactory and has become well established. Generally, speaking, leys do, however, deteriorate considerably towards the end of their life.

Valuing leys accurately at face value is a difficult matter. In the past, Valuers have tended to put rather low values on good leys which have considerable unexpired life and productivity. The value can be based on the market value of the pasture, whether for mowing or grazing, taking into account factors 5.2(*a*)–(*g*) above, with deductions made for the rental value (in accordance with Section 12 of the 1986 Act) and also any costs that would be incurred by way of application of fertilisers, sprays, pasture topping and, where used for mowing, the harvesting costs. This will produce a value for a season or year but percentage additions should be made for any further years of life the pasture is estimated to have.

An alternative basis possibly more accurate is to depreciate the cost of establishing the ley, on an annual basis, over its term, less any costs of spraying weeds or pasture topping, but making adjustments for factors (*a*) to (*g*) above.

Example Valuations

(i) *4 year ley* (at start of 2nd year valuation made at Candlemas Day)

		per Acre		*per Ha*
Cost of comparable grazing:				
Summer grazing (7 months)		£100.00		£247.00
Winter grazing (5 months)		20.00		49.00
Expense incurred supervising stock at away grazing say		5.00		12.00
Gross Cost		£125.00		£308.00
Less Deductions				
Rent per annum	£45.00		£111.00	
Fertilisers (3 applications of 20:10:10 compounds and 34.5% N)	£38.00		£94.00	
Spraying/Pasture Topping	5.00	£88.00	12.00	£217.00
Net value of ley in second year		£37.00		£91.00
Add for year 3—50%		18.50		45.00
Add for year 4—25%		9.25		22.50
Value of ley		£64.75		£158.50
	SAY	£65.00	SAY	£160.00

(ii) *Value of same ley in its last year* (Ley due up for Winter Wheat in Autumn. Valuation as at Candlemas)

Cost of comparable grazing (7 months) (now a poorer face)		£70.00		£173.00

Expenses incurred supervising stock at away grazing say		4.00		10.00
		£74.00		£183.00
Less Deductions				
Rent 7/12 × £45	£26.00		£64.00	
Fertilisers	35.00	61.00	86.00	£150.00
Net value of ley		£13.00		£33.00

Note: Some leys, near the end of their life, may have little or no value.

(iii) *Value of same ley (4 year) on depreciating basis (start of second year)*

Cost of establishing ley	£80.00 per acre
Less: depreciation for 1 year	20.00per acre
Value	£60.00 per acre

This is on the basis on no adjustments being made in respect of (*a*)–(*g*) above.

5.3. PRACTICE NOTES

(*a*) *the valuation must ignore any sod value*

(*b*) if the tenancy agreement schedules land as arable and it is in pasture at the termination date, the pasture is classed as a Tenants Pasture is the subject of a tenants outgoing claim (unless there is a written agreement that the Tenant cannot claim).

(*c*) if there is no written tenancy agreement in existence and the outgoing Tenant can prove, by producing his ingoing valuation inventory that he paid for any pasture on entry, or that any field was then arable and is now pasture, it can be the subject of a claim by the outgoer.

(*d*) if the pasture was ploughed up during and after the 1939/45 war, on Agricultural Executive Committee orders, and such written orders *can be produced*, any pasture subsequently laid down is a Tenant's Pasture and can be claimed for.

(*e*) valuing Tenant's Pastures is normally a somewhat contentious matter between Valuers, who frequently have widely differing opinions on the value of any given pasture.

(*f*) In making calculations of value for a season or year, do not forget to add for any further unexpired life the pasture will have after this period has expired.

(*g*) Valuers should carefully note the presence of weeds e.g. thistles, docks and dandelions, daisy etc. weed grasses e.g. agrostis and couch grass in pastures. These considerably reduce the value of pastures and are difficult to control.

CHAPTER SIX

Unexhausted Value of Feeding Stuffs Consumed
Unexhausted Value of Fertilisers Applied
Unexhausted Value of Lime Applied

6.1 The regulations relating to these are contained in *The Agriculture (Calculation of Value for Compensation) Regulations S.I. No. 809 1978* which are known as the 'principal regulations'. The principal regulations regarding this type of compensation have been amended by:

The Agriculture (Calculation of Value of Compensation) (Amendment) Regulations S.I. No. 822 1981. This S.I. contains the current table for depreciating lime.

The Agriculture (Calculation of Value for Compensation) (Amendment) Regulations S.I. 1475 1983. These regulations, operative as from 14th November 1983, replaced the previous compensation tables for unexhausted value of phosphate, potash, the value of purchased farmyard manure and the unexhausted manurial values of feeding stuffs consumed. However, note the provisions contained in the 'principal' regulations.

6.2. UNEXHAUSTED VALUE OF FEEDING STUFFS CONSUMED

6.2.1. Compensation tables are currently found in S.I. No. 1475 1983. Compensation is payable in respect of corn consumed on the holding (whether produced there or not) or any of the other purchased feeding stuffs listed in the tables that have been purchased and fed to cattle, sheep, horses, pigs and poultry on the holding. This includes *purchased* hay, straw etc., and any of the other items, in all 41, which are listed. Note, however, that silage or haylage are not included in this list.

6.2.2. Compensation is payable where no crop has been taken from the resultant farmyard manure or slurry made and on a diminishing scale for crops taken, both after one and two growing seasons. Different rates apply where the crop is fed on the holding and where the product of the manuring, e.g. hay sold off, is removed from the holding.

6.2.3 All the tables existing in the previous S.I.'s are now superseded by two tables—one for cattle, sheep and pigs (strangely horses are not mentioned here but reference is made to them on page 4 of the S.I.) and the other for poultry both on an open slurry basis. This does not necessarily mean that slurry has to be made as the compensation calculated on the tables is to be adjusted in accordance with Table 6 which deals with whether the manure is kept in a closed slurry pit (i.e. under slats or in a covered container) or whether it is farmyard manure stored under varying conditions, or whether the feeding stuffs has been fed directly on the land.

6.2.4. When calculating a claim, enquiry has to be made as to how the farmyard manure or slurry is dealt with, as this has considerable bearing on the correct calculation of the compensation payable. It is quite common for some farmers:

(*a*) to put their stock in strawed yards in early October and cart out the resultant F.Y.M. to heap in a field the following spring or summer. It is then spread in the autumn or early winter for the benefit of the following year's crop.
(*b*) in hilly and marginal land areas, in particular, to spread the previous winter's F.Y.M. immediately the following spring, on grassland, to produce hay crops.
(*c*) to spread slurry on the land two or three times in the winter and early spring.
(*d*) to leave slurry to solidify in the lagoon, or store, until the summer and then deal with it as (*a*) above.

6.2.5 Enquiry as to how the manure or slurry is dealt with, will result in the correct information being obtained for the proper calculation of the claim.

6.2.6 Item 41 of Table 5(a) and 21 of Table 5(b) refers to the percentage protein (or albuminoids) in compounded cake e.g. 18% albuminoids in 1 tonne of cattle cake consumed, no crop off, is 18 × 20p = £3.60 per tonne consumed.

6.2.7 *Specimen Claims*

(i) 10 Tonnes Sugar Beet Nuts fed to beef cattle on slats. No crop off

10 Tonnes × 309 =	£30.90	
Conditions Ideal—Add 30%	£9.27	£40.17

(ii) 20 tonnes milk nuts 16% alb. fed to dairy cattle. After one growing season—hay crop sold off. Farmyard manure made—average conditions.

20 tonnes × (8p × 16%) =		
20 × 128 =	£25.60	
Add 40% (Table 6)	£10.24	£35.84

(iii) 30 tonnes homegrown seed hay fed to outwintered cattle. After two growing seasons. Product fed on holding. No claim since homegrown hay is not eligible.
Note: If the hay was purchased and brought onto the holding the claim would be:

30 Tonnes × 84p =	£25.20	
Add 35% (Table 6)	£8.82	£34.02

(iv) 30 tonnes homegrown barley fed to pigs. Farmyard manure made. Ideal conditions. No crop off.

30 Tonnes × 204p =	£61.20	
Add (Table 6) 50%	£30.60	£91.80

6.3 UNEXHAUSTED VALUE OF FERTILISERS APPLIED

6.3.1 Nitrogen

No compensation is payable after one crop has been taken from the land.

6.3.2 Phosphoric Acid (P_2O_5)—Phosphates

Compensation is payable in accordance with Table 2 Part II of S.I. 1475 of 1983.

Reference should be made to S.I. No. 809 1978 (Pages 6 & 7) which states:

(*a*) where the phosphatic fertiliser contains less than 1/10 of its total phosphoric acid content in an insoluble form, the total phosphoric acid content shall be treated as soluble.

(*b*) where it is a fertiliser (other than a fertiliser specified and applied as in 2(a); 2(b)(i) or 3(a) in Table 2) containing more than $\frac{1}{10}$ of its total in soluble form, the value shall be restricted to the soluble form only. The permanent grassland referred to has to be established for 5 or more years.

Different rates apply, dependent on the type of fertiliser applied, as detailed in the table and care should be taken to find out exactly what type of phosphate is used. Regarding Item 4 of Table 2 Calcined calcium aluminium phosphate, to substantiate a claim here, it is necessary to get the soil, where it was applied, analysed. Obviously, if the phosphate level is good, compensation would be payable, by negotiation, of the amount considered fair.

(*c*) where the phosphate applied is rock phosphate different rates apply where this is 'soft' and these rates have to be adjusted having regard to the mount of the mean annual excess winter rainfall (M.A.E.W.R.) (different rates apply when the rainfall is above and below 450 mm and also when applied to *permanent* grassland and other crops.)

(*d*) *Specimen Claims*

(i) 10 tonnes Phosphate 23% (20.6% Sol. 2.4% Insol.) applied to pasture. After one growing season.

Calculation is 10 tonnes × (20.6 × 158p)
= 10 × £32.54 = £325.40

(ii) 5 tonnes soft ground rock Phosphate 21% (19.9% Sol. 1.1% Insol.) applied to forage crop in area of 400 mm M.A.E.W.R. After 2 growing seasons.

Claim NIL

(iii) Same quantity and analysis as (ii) applied at same date to permanent pasture in area of less than 450 mm M.A.E.W.R. Calculation is:

5 tonnes × (21 × 79p)(2 GS)
5 × 1659p = £82.95

6.3.3 Potash (K_2O)

Compensation is payable in accordance with the rates shown in Table 3 of S.I. 1475 of 1983. It is restricted to two growing seasons and then payable *only* in respect of arable crops (except forage crops), root crops (where the tops are *left* on the land e.g. sugarbeet) and also leys, permanent grassland and forage crops, grazed or the product cut and fed on the holding.

(*a*) Nothing is payable in respect of potash applied to potatoes, roots, forage crops and applied to leys and permanent grassland where the product (e.g. hay or silage) is *removed* from the holding.

(*b*) The Valuer should read Paragraph 6(2)(c) of S.I. No. 809 of 1978 which inter-alia states that where the Holding is mainly horticultural, the value calculated should be in accordance with Item 1 of Table 3 (now S.I. 1983 No. 1475).

(*c*) *Specimen Claims*

(i) 10 tonnes Potash 20% K_2O applied to sugar beet crop (tops part folded off, part ploughed in). After one growing season. Calculation is:
10 tonnes × (20 × 92p)
10 × 1840p = £184.00

(ii) 10 tonnes Potash (20% K_2O) applied to potato field (after one growing season)
No claim.

6.3.4 Compounds

Most fertilisers used are compounds of Nitrogen, Phosphate & Potash.

After *one* crop has been taken, no compensation is payable in respect of nitrogen but the compensation payable in respect of phosphates and potash shall be in accordance with the Tables in S.I. 1475 of 1983.

(*a*) *Specimen Claim*

(i) 10 tonnes Fisons 52—20.10% (1% Insol.).10 (the analysis is always shown in the order Nitrogen:Phosphate:Potash) applied to permanent grazing meadows. Claim after one growing season is:

Nitrogen	Nil	
P_2O_5 10 tonnes × (10 × 158p) =	£158	
K_2O 10 tonnes × (10 × 92p) =	92	
Total		£250

(ii) 10 tonnes Fisons Maincrop Potato 13.13(1.2 Insol).20 applied to maincrop potatoes—after 2 growing seasons. M.A.E.W.R. 460 mm.
Claim is:

Nitrogen	Nil	
P_2O_5 10 tonnes × (13 × 79p) =	£102.70	
K_2O_5	Nil	
Total		£102.70

6.4. FARMYARD MANURE

Compensation for home produced farmyard manure is calculated through the feeding stuffs consumed and nothing, otherwise, is paid, except for the labour involved. Note maximum compensation is for 50 tonnes/ha [20T/Ac] Cattle, Horse & Pig Manure; 18 tonnes/ha [7.28T/Ac] Deep Litter Poultry Manure & 12½ tonnes/ha [5T/Ac] Broiler Poultry Manure.

6.4.1 If farmyard manure (this includes poultry and horse manure) is brought onto the holding and no crop has been taken since the manure was applied, the value shall be for the cost of delivery and application where no payment was made for the manure. In this case the value in subsequent years, after the first growing season, is half, and after the second growing season, is one quarter, of the cost of delivery and application (S.I. 1978 No. 809). No compensation is payable in respect of purchased manure in the last year of tenancy, after the last crop was removed, unless the landlord gave written consent for such application.

6.4.2 *Where payment is made for the manure*, the claim shall be for the cost of the manure, its delivery and application where no crop is taken, and after one and two growing seasons, this shall be reduced to one half and one quarter respectively, *provided* the value of the manure specified in Table 4 of S.I. 1983 No. 1475 shall not exceed the figures given in that table.

6.4.3 *Specimen Claims*

(*a*) 100 tonnes purchased broiler manure, after 2 growing seasons. Cost including spreading £1,560

Claim – ¼ Cost = £390

(*b*) 80 tonnes pig farmyard manure (free for fetching) after one growing season.
Cost of collection and spreading £760

Value after 1 season half = £380

6.5 UNEXHAUSTED VALUE OF LIME APPLIED

Compensation provisions here are contained in S.I. No. 822 1981 Page 3. The calculation appears somewhat complicated, but is not so, providing the following are ascertained:

(*a*) the amount of the mean annual excess winter rainfall (definition given in S.I. 822).
(*b*) the amount of nitrogen applied annually to the land concerned. If the land is in permanent pasture and long term leys or in arable rotation (with short term leys) and *more* than 250kg of nitrogen per ha is applied annually, the cost of the lime applied (this includes cost of delivery and application) can be depreciated, dependent on M.A. Excess Winter Rainfall from 4 years (where the rainfall exceeds 500 mm M.A.E.W.R.) up to 8 years where it is less than 250 mm.

If nitrogen applied is *up to 250 kg* per ha annually, and the land concerned comprises permanent pasture or long term leys, the cost of the lime applied is depreciated over 9 years where the M.A.E.W.R. is less than 250 mm and over 5 years where it is more than 500 mm.

In cases when the rainfall is between 250 and 500 mm the depreciation is over 7 years.

Note 1: For period of *one year* following the application of the lime the value shall be its cost applied.

Note 2:

(*a*) the cost shall not be regarded as reasonable to the extent that it exceeds the higher of:
  (i) the cost, at time of application, of the limestone or chalk applied (whichever is cheaper) which would have been used in the application to the land of calcium oxide at a rate $7\frac{1}{2}$ tonnes per ha and
  (ii) the cost of lime recommended in scientific advices relating to the condition of the soil.
(*b*) Cost includes delivery and application.

6.5.1 *Specimen Claims*

(*a*) 100 tonnes lime applied to long term pasture after 2 growing seasons. Cost including delivery and spreading was £1050.

450 mm M.A.E.W.R. 150 kg N per ha applied annually.
Claim will be:
Cost depreciated for 2 years = $\frac{2}{7}$ off cost £1050 = £750.

(*b*) 58 tonnes lime applied to field in regular arable cropping after 4 growing seasons. Cost including delivery and spreading was £500. Average of 350 kg N per ha applied annually. 210 mm M.A.E.W.R.
Claim will be:
Cost depreciated for 8 years = $\frac{4}{8}$ off cost £500 = £250.

CHAPTER SEVEN
Sod Fertility Claims

7.1. This type of claim was introduced by S.I. 1978 No. 809 and this has since been amended by S.I. 1980 No. 751 and later by S.I. 1983 No. 1475.

7.2. This claim is intended to compensate an outgoing tenant for increased fertility resulting from leaving more pasture (and thus with more inherent fertility) than he is required to, under the terms of the tenancy agreement.

7.3 For a claim to succeed a written notice must be served on the landlord, prior to the termination of the tenancy, stating that the provisions of Part II of the Eighth Schedule of the 1986 Act under the provisions of S.I. 1978 No. 742 shall apply to him as regards such claim.

7.4 A stipulation in S.I. No. 1475 states that the holding must be in an area *'where arable crops can be grown in an unbroken series of not less than 6 years and it is reasonable that they should be grown on the holding, of part thereof'.*

7.5 The calculation is very complicated and a claim will not succeed unless very detailed and reliable records can be produced, for several past years, of the cropping of the various fields forming the holding.

7.6 Explanation of matters referred to in Paragraph 12 of S.I. 809 1978 are as follows:

(a) Accepted Proportion of Leys

This is the area specified in the tenancy agreement of leys or new seeds to be left at the tenancy termination. If there is no agreement or it is silent on the matter it is the area representing the proportion which the total area of leys on the holding would, taking into account the capability of the holding, be expected to bear to the area of the holding, excluding permanent pasture.

(b) Excess Area of Leys

This is the average of the *total* area of leys at the termination date, and one and two years prior to that date *less* the accepted area required for the fertility of the holdings. The surplus is the excess area which may be considered for compensation.

(c) Qualifying Ley

This is an established ley (3 or more years old) or what was a former established ley, that is one which was three years old or older before ploughing or destroying.

Note: That permanent pasture is excluded, as are leys laid down at the expense of the landlord, without payment, by the tenant or any previous tenant.

(d) Calculation of Compensation

The residual value of the sod of the excess qualifying leys is calculated (but subject to the amendments contained in Paragraph 3 (2)(d)(i) & (ii) of S.I. 1475 1983) as follows:

(i) in respect of continuously (3 or more growing seasons) maintained leys £24 per ha if the crop has been cut and removed in the last growing season or £40 per ha if grazed only in the last season;

(ii) regarding continuously maintained leys, the value shall be increased by £8 per ha for each additional growing season over three, but subject to a maximum of £48 per ha where the crop was cut and removed or £64 per ha where grazed only in last season.

(iii) in respect of any *former* ley where the first crop which has been sown in the last growing season before termination of tenancy has not been removed from the ground, the value shall be the value specified in (i) and (ii) above according to the period for which the ley had been established before ploughing up or destroying and to whether the herbage was cut and removed, or grazed only, in the last season before ploughing etc.

(iv) in any former ley to which (iii) does not apply

(i)(aa) if only one arable crop was taken following ploughing etc., value is 2/3 of that specified in (i) and (ii) above;

(bb) if two arable crops taken, value is ⅓ according in each case to the period for which the ley was established and to whether the ley was cut and removed or grazed in last season before ploughing;

(ii) if more than two arable crops taken since ploughing—value is nil.

Where tenant is entitled to compensation for *both tenants pasture (under Paragraph 10 S.I. 809 1978) and sod value under (d)(i) and if applicable (d)(ii) above* the aggregate of the respective values per hectare, taken together, shall not exceed £148 per ha.

An example of a claim on a 317 acre (128.3 ha) holding is as follows:

Sod Value Claims

Arable acreage on Schedule of Agreement: 239.522 ares (96.97 ha)
Area to be left as Leys on Agreement: $\frac{2}{5}$th = 38.78 ha

		Leys			*Year laid*	
O.S.	*No.*	*Acre*	*Ha*	*Description*	*down*	*Last year*
Pt.	54	6.00	2.42	Long Ley	1981	Grazed
	54	16.569	6.70	Long Ley	1981	Grazed
	53	17.143	6.94	Long Ley	1981	Grazed
	619	11.691	4.73	Long Ley	1980	Mown
	651	17.374	7.03	Long Ley	1980	Mown
	654	14.043	5.68	Long Ley	8 years ago	Grazed
	58	9.370	3.79	Long Ley	8 years ago	Grazed
Pt.	149	12.503	5.06	Long Ley	8 years ago	Grazed
	221	8.649	3.50	Long Ley	8 years ago	Grazed

Cropping after Leys

220	18.341	7.42	Barley 1989	Wheat 1988 Previous 4 years Ley Last mown 1987
617	9.368	3.79	Roots 1989	Wheat 1988 Ley 1982–87 Mown 1987
604	7.263	2.94	Barley 1989	After 4 years Ley sown 1985 Mown 1987
652	8.202	3.32	Barley 1989	After 6 years Ley sown 1983 Mown 1988

653	4.985	2.01	Barley 1989	Barley 1988 After 4 years Ley sown 1985 Grazed 1987
656	15.094	6.11	Wheat 1989	Pasture for 10 years Grazed 1988

Sequence of Leys

O.S. No.	*Acres*	*Ha*	*1982*	*1983*	*1984*	*1985*	*1986*	*1987*	*1988*	*1989*
Pt. 54	6	2.42	L	L	L	L	L	L	L	L(G)
Pt. 54	16.569	6.70	L	L	L	L	L	L	L	L(G)
53	17.143	6.94	L	L	L	L	L	L	L	L(G)
619	11.691	4.73	L	L	L	L	L	L	L	L(M)
651	17.374	7.03	L	L	L	L	L	L	L	L(M)
654	14,043	5.68	L	L	L	L	L	L	L	L(G)
58	9.370	3.79	L	L	L	L	L	L	L	L(G)
Pt. 149	12.503	506	L	L	L	L	L	L	L	L(G)
221	8.649	3.50	L	L	L	L	L	L	L	L(G)
220	18.341	7.42			L	L	L	L(M)	W	B
617	9.368	3.79	L	L	L	L	L	L(M)	W	R
604	7.263	2.94				L	L	L(M)	L	B
652	8.202	3.32		L	L	L	L	L	L(M)	B
653	4.985	2.01			L	L	L	L(G)	B	B
656	15.094	6.11	L	L	L	L	L	L	L(G)	W

All 'qualifying leys' as are continuously maintained leys or former leys

L = Ley	G = Grazed	R = Roots
M = Mown	B = Barley	W = Wheat

Eligible Leys in Excess of Area reasonably required for Fertility

1. Area of 'Qualifying leys' in each of last 3 years, including year sown all as Spring sown

O.S. No.	*1989*	*1988*	*1987*
Pt. 54	2.42	2.42	2.42
Pt. 54	6.70	6.70	6.70
53	6.94	6.94	6.94
619	4.73	4.73	4.73
651	7.03	7.03	7.03
654	5.68	5.68	5.68
58	3.79	3.79	3.79
Pt. 149	5.06	5.06	5.06
221	3.50	3.50	3.50

220			7.42
617			3.79
604		2.94	2.94
652		3.32	3.32
653			2.01
656		6.11	6.11
	45.85	58.22	71.44
		Mean	58.50 ha

Less

2. Total area of leys to be left sown, reasonably required for fertility of holding is $\frac{2}{5}$ of arable area of 96.9 ha = 38.78 ha.

3. *Equals*
   58.50 ha − 38.78 ha = 19.72 ha Compensationable

Valuation Summary

GS = No. of growing seasons including year sown
G = Grazed
M = Mown

O.S. No.	*Ha*	*GS*	*G/M*	*Years cropped after grass*		*S.F.V. per Ha*	
Pt. 54	2.42	3	G			£40.00	
Pt. 54	6.70	3	G			40.00	
53	6.94	3	G			40.00	
619	4.73	4	M		(24 + 8)	32.00	
651	7.03	4	M		,,	32.00	
654	5.68	8	G			64.00	(max)
58	3.79	8	G			64.00	(max)
Pt. 149	5.06	8	G			64.00	(max)
221	3.50	8	G			64.00	(max)
220	7.42	6	M	2	(24 × $\frac{1}{3}$)	8.00	
617	3.79	6	M	2	,,	8.00	
604	2.94	5	M	2	,,	8.00	
652	3.32	7	M	1	(24 × $\frac{2}{3}$)	16.00	
653	2.01	6	G	2	(24 × $\frac{1}{3}$)	8.00	
656	6.11	8	G	1	(24 × $\frac{2}{3}$)	16.00	

Fields selected for maximum compensation

654, 58, Pt. 149, 221	18.03 ha @ £64 =	£1,153.92
Pt. 54	1.69 ha @	67.60
Total	19.72 ha	
Total compensation		£1,221.52

Note: The S.F.V. per ha is prescribed in S.I. no. 809, 1978, as amended by S.I. No. 751, 1980, as further amended by S.I. No. 1475, 1983.

CHAPTER EIGHT

Tenant's Improvements

8.1 This chapter deals with Long Term Improvements e.g. erection of buildings, reclamation of land, making of roads, provision of sheep-dipping accommodation for which Landlords consent is required or approval of the Agricultural Lands Tribunal.

Distinction should be made between Old Long Term Improvements—those carried out *before* 1st March 1948 and those called New Long Term Improvements—those carried out *after* 1st March 1948.

8.2 *Old Long Term Improvements* (Schedule 9 Part 1, 1986 Act)—these are major improvement works, such as building erection and are still frequently encountered despite being over 40 years old. If the consent given was an open one and did not specify agreed terms of compensation or is under custom, then Paragraph 2 of Schedule 9 of the agricultural Holdings Act 1986 provides that the compensation payable shall be *'an amount equal to the increase attributable to the improvement in the value of the Agricultural Holding as a holding, having regard to the character and situation of the holding and the average requirements of tenants reasonably skilled in husbandry'*.

8.3 *Long Term New Improvements* (Part II Schedule—7, 1986 Act) These again, are major improvements, broadly in line with Old Long Term Improvements, but not entirely so).

8.3.1 These, improvements again, require the Landlord's or the Agricultural Land Tribunal's consent. Many Landlords and their agents give consent over a set write off period, often over 10 or 15 years, (but in the case of specialised buildings such as pig or dairy units, sometimes less) with the net cost (after grant) being written down to nil. Any write off should be down to say £1—if only to recognise the tenant's ownership of the improvement.

8.3.2 It is considered by many valuers that 10–15 years' write off period is far too low for substantial buildings such as a covered yard or Dutch Barn which can reasonably be expected, provided it is well built, designed and properly maintained to last 50 years or so.

8.3.3 If the Landlord gives an open unrestricted consent (which, incidentally, can also be given after the improvement has been effected) compensation provisions on the termination of the tenancy are governed by Section 66 of the 1986 Act, and the *amount of compensation shall be an amount equal to the increase attributable to the improvement in the value of the holding as a holding*, i.e. a let farm, and *regard has to be had to the character and situation of the holding and the average requirements of tenants reasonably skilled in husbandry.* This infers that the improvement must be suitable for the holding e.g. it could be inappropriate to have an expensive dairy set up on a hill farm, or possibly an expensive grain store on a wet farm which is more suitable for grazing than for arable production. The usual method of valuation is to assess the increase in the rental value of the holding arising from the improvement and capitalising this at a suitable rate of interest for its remaining estimated life and thus produce the increased value of the holding. Regard must be had, of the likely life of the improvement, e.g. an electrical wiring installation will probably need renewing in 15 years' time, whilst a Dutch Barn may last 50 years. Examples:

(a) Covered cattle yard 90 ft × 50 ft erected by tenant 4 years ago. Net cost after grant £22,000. Farm 150 acres in area and short of modern buildings.

This is a good substantial modern building that is needed on the farm and can be put to almost any use, and is expected to last 30 years. It is considered that the erection of this building has increased the rental value of the holding by £1,500 per annum. The claim will therefore be:

Increased rental value (net)	£1,500
Y.P. 26 years @ 12%	7.9
Increase in value of holding	£11,850

(b) A tenant has reclaimed an area of 20 acres rough land, including providing certain drainage and fencing works. It is now good pasture with a rental value of £30 per acre which is expected to continue (subject to inflation or deflation) in perpetuity. Its unimproved value would have been £4 per acre. The claim would be calculated viz:

Current Rental Value 20 acres @ £30	£600
Less: Unimproved Rental Value	£80
	£ 520
Y.P. 20 years @ 12%	7.47
	£3,884 say £3,900

8.4 *Note*: The relevant section of the 1986 Act should be studied carefully. Sometimes, tenants have carried out Long Term Improvements without Landlord's consent and cannot, therefore, obtain compensation. However, their position my not be totally lost as, by giving the requisite notice, the items may be treated as *Tenant's Fixtures*, and the improvement can be removed or be taken over by the landlord, should he elect to do so.

8.5 Note that application of purchased manures, liming and consumption on the holding of corn (whether produced there or not) and of cake or other feeding stuff not produced on the holding are *improvements* and not Tenant Right. Landlord's consent for their use, however, is not needed.

CHAPTER NINE
Tenant's Fixtures

9.1 Section 10 of the Agricultural Holdings Act 1986 gives the Tenant the right to remove buildings, engines, machinery, fencing or other fixtures of whatever description affixed to an agricultural holding by the Tenant, provided the fixture is not affixed in pursuance of some obligation to do so. The Tenant may remove the fixture at any time during the tenancy or before the expiry of two months from the termination of the tenancy, providing he has paid all rent owing and satisfied all his Tenant's obligations and also has, at least one month before the exercise of the right and the termination of the tenancy, served on the landlord a written notice of his intention to remove the fixture or building. The Landlord can serve on the Tenant a written Counter-Notice before the expiration of the Notice electing to purchase the fixture etc., and the measure of compensation payable by the Landlord is the *'fair value of the fixture or building to an incoming Tenant'*. Any dispute as to the amount of compensation shall be determined by arbitration.

9.2 This simply means that the compensation is what an incomer would be expected to give for the fixture or what it is worth to him.
Factors to take into account in assessing this compensation are:

(i) The current net cost of providing the fixture.

(ii) The existing condition of the fixture, taking into account the cost of any repairs necessary.

(iii) Whether the fixture is a modern one or not—perhaps it may be outdated e.g. an abreast milking parlour may be considered outdated, or a building erected may have too low a headroom.

(iv) The expectation of the fixtures remaining useful life.

(v) Is the fixture of value to an incoming Tenant or not? e.g. a Landlord may have elected to take over a piggery erected by the outgoing Tenant, but the incoming Tenant does not intend keeping pigs and to the incomer may it only be of use for his ewe flock at lambing time. It should, however, be borne in mind that in this example, the valuation is between the Outgoing Tenant and Landlord, and since the Landlord has elected to take to the fixture, it presumably has value to the Incomer. There may be other factors to be considered, but the valuation of fixtures, as in the case of Tenant's improvements, is difficult and often arguable.

9.3 *Examples* (Note: In each case below Counter-Notices are deemed to have been served to purchase the fixtures, all of which will be of value to the incoming Tenant.)

(i) The outgoing Tenant erected, 5 years ago a 6/12 herringbone parlour and adjacent dairy. He provided all the milking equipment including 1,350 litre bulk milk tank, milking equipment etc. The total (net) cost of the work was £7,500 for the buildings and £4,000 for the equipment. Repairs estimated to cost £1,000 are necessary to the buildings and £500 for the equipment. The buildings are estimated to last a further 10 years and the equipment 5 years, when both will need replacing at current estimated net cost, respectively of £15,000 and £8,000. It is suggested that the claim be dealt with as follows:

Building			
Cost (net) of replacement		£15,000	
Less: $\frac{1}{3}$ life expired	£5,000		
Repairs	1,000	6,000	
Net value			£9,000
Equipment			
Cost of replacement		£8,000	
Less: $\frac{1}{2}$ life expired	£4,000		
Repairs	500	4,500	3,500
Net value			£12,500

Note: The compensation is higher than the original cost, but to re-new the whole would now cost £23,000 so the figure arrived at, taking into account inflation, is probably fair.

(ii) Pig netting fencing of 400 m run and one $4\frac{1}{2}$ m wide gate erected in O.S. No 12 five years ago. Repairs estimated to cost £50 are necessary. The whole is expected to have a life of 15 years from date of erection. Current net cost of the work is estimated to be £750.

Estimated replacement costs		£750	
Less: $\frac{5}{15}$ life expired	£250		
Repairs	50	300	
Net value			£450

(iii) A steel and asbestos implement shed 60 ft × 30 ft erected 20 years ago. Eaves height 8 ft (which is considered too low for modern equipment). It has a further expected life of 25 years but guttering requires replacement, steelwork re-painted and doors repairing at a cost of £750. Current replacement cost (net) is estimated to be £7,000. Claim would be:

Replacement cost		£7,000
Less: Repairs	£750	
$\frac{20}{45}$ life expired	3,111	3,861
		£3,139
but since the building is too low for modern requirements deduct $\frac{1}{3}$		£1,046
Net value		£2,093

Say—£2,000

(iv) A sheep dip, foot bath and handling pens constructed five years ago. The replacement cost today would be £5,000 (net). Repairs estimated to cost £500 are necessary. The set up has an estimated unexpired life of five years and will be of use to the incoming Tenant. The claim would be:

Replacement cost		£5,000
Less: Repairs	£500	
$\frac{1}{2}$ of life expired	2,500	3,000
Net value		£2,000

CHAPTER TEN

Offgoing Crops (or Away Going Crop)

10.1 Certain old tenancy agreements sometimes provide for an acreage of offgoing crop of winter wheat. These tenancies were always Candlemas (2nd February) or Lady Day (25th March) and the principle was that the outgoing Tenant should retain the crop he planted and that he should be allowed back on the farm to harvest his crop, despite the tenancy having terminated. This proved inconvenient and it became standard practice to value the crop and commute the profit into a cash payment. In effect, the offgoing crop right is the profit on a crop of wheat, less risk (it may become diseased etc.). At a meeting of a local Agricultural Valuers Association, held in 1957, the following matters and basis of valuation were agreed:

(i) that where the right is established and unless the agreement states to the contrary, the offgoing crop right should be $\frac{1}{4}$ of the arable area.

(ii) unless the agreement decrees otherwise, the crop must be wheat.

(iii) the crop must be planted by Candlemas Day (2nd February).

(iv) the valuation is the estimated value of the crop (the straw is left free to the new tenant) deducting harvesting costs and a percentage for risk.

10.2 *Example Valuation of Offgoing Crop Right*

			per Acre		*per ha*
Yield:	$2\frac{1}{2}$ tonnes @ £110		£275		
	6.17 tonnes @ £110				£678
Less:	Spraying herbicide	£15.00		£37.00	
	Harvesting	25.00		62.00	
	Risk 25% say	70.00	110	173.00	272
Profit i.e. Value of Right			£165		£406

Example of Claim at the Termination of a Tenancy

(2nd February 1990)

AGRICULTURAL HOLDINGS ACT, 1986

Re: The Holding known as
WEST FARM, METCHINGFIELD, HEREFORD

To: THE WALFORD ESTATE COMPANY (Landlords)
ESTATE OFFICE,
METCHINGFIELD,
HEREFORD.

I HEREBY GIVE YOU NOTICE pursuant to Section 83(2) of the above Act of my intention to make against you certain claims arising out of the termination of the tenancy of the above holding, the nature of which claims is set out in the Schedule hereto.

I also hereby give you notice, under paragraph II of Schedule 1 of the Agriculture Act 1986, that it is my intention to make against you the claim specified in the Schedule hereto, being a claim for compensation in respect of Milk Quota, which claim arises under para. 1 of Schedule 1 of the Said Act on the termination of my tenancy of the said holding.

THE SCHEDULE

Claims made under Clause VII of the Tenancy Agreement dated 4th April 1972 and the Surrender Agreement dated 20th September, 1989 and principally in accordance with S.I. 1978 No 809 S.I. 1981 No. 822 and S.I. 1983 No. 1475

1. PRODUCE

(a) BARLEY STRAW

Long Barn

84.6 tonnes 1989 Baled Straw @ £25	£2,115.00

Home Barn

25 Big Bales 1988 Straw Av. weight 230 kg 5.75 tonnes @ £25	£ 143.75

(b) HAY

Long Barn

25.4 tonnes 1989 Seeds Hay @ £70	£1,778.00

Covered Yard

42 Big Bales 1989 Meadow Hay Av. weight 510 kg—21.42 tones @ £70	£1,499.40

(c) SILAGE

Clamp at Homestead

Measurements 25 m × 13 m × 2 m
Sample Silage £18 per tonne

Value of Silage per tonne is £20.96

$$\text{kg per m}^3 = \frac{6500}{26} + 400 = 650$$

$\therefore$ 25 m × 13 m × 2 m = 650 m³ @ 650 kg
= 422.5 tonnes @ £20.96 = £8,855.60

£14,391.75

2. GROWING CROPS

(a) O.S. Pt. 602 8.79 ha—**Sonja Winter Barley**

8.5 ha Planted

	£ p		
1 Plough	28.15		
3 Disc Harrow	24.00		
1 Drill	9.40		
1 Harrow	4.30		
1 Fertiliser Spinner	5.80		
1 P/E Spray	6.40		
8.5 ha @	£78.05	663.42	
Seed			
30 × 50 kg @ £11 per 50 kg (ex. M.S.F. A/C 10)		330.00	
Fertilisers			
42 × 50 kg 9:25:25 (Ex M.S.F. A/C 11) @ 140 per tonne		294.00	
Sprays (Ex M.S.F. A/C 12)		45.00	
Enhancement Value			
£36 per ha × 8.5		306.00	£1,638.42

(b) O.S. No. (pt) 652 3.32 ha
O.S. No. 604 2.97 ha
O.S. No. 220 7.42 ha
O.S. No. 218 6.87 ha
20.58 ha

AVALON WINTER WHEAT 20 ha Planted

	£ p
1 Plough	28.15
2 Disc Harrows	16.00
1 Roterra	18.05
1 Combine Drill	16.40
1 Harrow	4.30

1 P/E Spray	6.40		
20 ha @	£89.30	1,786.00	
Seed			
74 × 50 kg @ £11.25 per 50 kg (ex M.S.F. A/C 10)		832.50	
Fertilisers			
100 × 50 kg 9:25:25 @ £140 per tonne (Ex. M.S.F. A/C 11)		700.00	
Sprays (Ex. M.S.F. A/C 12)		105.88	
Enhancement Value			
£36 per ha × 20 ha		720.00	£4,144.38
(c) O.S. No. 618. 1 ha SWEDES			
(Planted 16th June 1989)			
	£ p		
1 Plough	28.15		
1 Disc Harrow	8.00		
1 Roll	10.35		
1 Harrow	4.30		
1 Drill	22.70		
1 Fertiliser Spinner	14.20		
1 ha @	87.70	87.70	
Seed (Ex. M.S.F. A/C 14)		24.70	
Fertilisers			
13 × 50 kg Compound @ £130 per tonne		84.50	£ 196.90
			£5,979.70

3. CULTIVATIONS

O.S. No. 653	2.01 ha	
O.S. No. 616	4.56 ha	
O.S. No. 832	5.25 ha	
O.S. No. 617	2.43 ha	
PLOUGHED		
14 ha @ £28.15		£ 394.10

4. LABOUR TO F.Y.M.

(a) Heap in O.S. 655		
22 m × 15 m × $\frac{2}{3}$		
220 m^3 @ 90p	198.00	
(b) Heap in O.S. 52		
11 m × 15 m × $\frac{2}{3}$ m		
110 m^3 90p	99.00	£ 297.00

5. TENANTS PASTURES				
(All scheduled Arable in Tenancy Agreement)				
(a) O.S. No. 54 9.10 ha				
4 year ley in last year				
9 ha @ £37			333.00	
(b) O.S. No. 53 6.94 ha				
4 year ley sown Aug 1988				
6.75 ha @ £125			843.75	
(c) O.S. No. 651 7.02 ha				
O.S. No. 654 6.68 ha				
13.70 ha				
Long term ley sown 1986				
13.5 ha @ £85			1,147.50	
(d) O.S. No. (pt) 145 5.06 ha				
4 year ley sown after winter				
Barley—Aug 1988				
5 ha @ £110			550.00	
(e) O.S. No. 58 3.79 ha				
O.S. No. 221 3.50 ha				
O.S. No. 619 4.72 ha				
12.01 ha				
Permanent Leys laid down in 1980				
11½ ha @ £62.50			718.75	
(f) O.S. 280 6.50 ha Young Seeds 3 Year Ley				
Sown August 1989 (not Grazed)				
1 Plough	28.15			
1 Roterra	18.05			
1 Drill (Broadcast)	5.45			
1 Fertiliser Spinner	5.70			
1 Roll (gang)	5.70			
6 ha @	63.15	378.90		
Seeds				
(Ex M.S.F. A/C 15)		360.00		
Fertilisers				
30 × 50 kg 20:10:10 @ £133 per tonne		199.50		
Enhancement Value				
6 ha @ £36		216.00	£1,154.40	£4,747.40
6. SOD VALUES				£467.84

7. RESIDUAL VALUE OF FEEDING STUFFS CONSUMED — £987.00

8. UNEXHAUSTED MANURIAL VALUES — £1,262.87

9. RESIDUE OF LIME APPLIED — £421.61

10. TENANTS IMPROVEMENTS — £3,260.00
    As already agreed

11. MILK QUOTA
    allocated quota in excess of standard quota;
    Tenants fraction of standard quota;
    Purchased quota of which Tenant has borne entire cost.
    As already agreed — £30,811.00

SUMMARY

1.	Produce	14,391.75
2.	Growing Crops	5,979.70
3.	Cultivations	394.10
4.	Labour to F.Y.M.	297.00
5.	Tenants Pasture	4,747.40
6.	Sod Values	467.84
7.	R.V.F.S.	987.60
8.	U.M.V.'s	1,262.87
9.	Residue of Lime	421.61
10.	Tenants improvements	3,260,00
11.	Milk quota	30,811.00
	TOTAL	£63,020.87

Dated this 30th day of January 1990

....

WILLIAMS, DAVIES AND REES,
Agricultural Valuers,
Dolanog,
Welshpool,
Powys.

Authorised Valuers Agents on Behalf of the Outgoing Tenant
Mr. J. Yeoman-Farmer

E. & O.E.

CHAPTER ELEVEN
Dilapidations

11.1 LIABILITY

The Landlord of an agricultural holding, can, under the provisions of Section 71 (1) of the Agricultural Holdings Act 1988, on the tenant quitting the holding, on the termination of the tenancy recover compensation in respect of dilapidations, or deterioration or damage to any part of the holding by the nonfulfilment by the tenant of his responsibility to farm the holding in accordance with the rules of good husbandry. These rules are laid down in Section 11 (1)–(3) of the Agriculture Act 1947 and briefly cover proper maintenance of pasture, keeping the land clean and in a good state of cultivation, keeping the farm properly stocked, maintaining efficient management of livestock, maintaining crops and livestock free from disease etc, protection and preservation of harvested crops and carrying out necessary maintenance and repair work.

The amount of the compensation recoverable shall be the cost, at the date of quitting, of making good the dilapidation, deterioration or damage (Section 71(2)).

However, the Landlord can, instead, of claiming under Section 71 (1) claim under the terms of a tenancy agreement (Section 71 (3)), which is the normal basis under which most dilapidation claims are made. Where a claim is made under the provisions of the tenancy agreement, Section 71 (4)(b) states that compensation shall *not* be claimed in respect of any one holding, under *both* such tenancy agreement *and* Section 71(1).

Section 71(5) provides that the compensation recoverable under Section 71(1) and 71(3) shall not exceed the amount (if any) by which the value of the landlords reversion in the holding is diminished owing to the dilapidation, deterioration or damage in question.

This latter overriding provision, which was introduced in 1984, has resulted in some dilapidation claims becoming very contentious.

11.2 Note the difference between dilapidations and 'general deterioration'. Section 72 of the 1986 Act gives the Landlord an additional remedy insofar as a specific claim for dilapidations does not effectively give him adequate compensation for his loss. Examples of the deterioration of a holding are allowing the farm to be severely neglected in almost every respect, including severe loss of soil fertility or serious infestation of diseases, weeds, etc., possibly resulting in lower yields for some period in the future. A claim for deterioration will not be effective unless at least a month's written notice of intention of claim has been given *before the termination of the tenancy*.

11.3 Any claim for dilapidations arising at the termination of a tenancy must be served within *two months of the termination date.* It is sufficient to give a notice specifying the nature of the claim and also under what term of the tenancy agreement, rules of good husbandry; model clauses etc., it arises. In practice a claim is lodged in detail and this also specifies the amount of each item of claim. In preparing each claim it is sound practice to specify the item of dilapidation with clear reference to where it exists e.g. in which O.S. No. or on what boundary between one O.S. No. and another. Each item of claim should be numbered (this is useful in subsequent negotiations), the basis of computing the claim should be shown together with the pricing mechanism e.g. 2 worker days @ £39.80 or 105 metres of overgrown hedge to cut and lay £2.61 per m (114½ yds @ £2.40).

11.4 The claim should be set out properly for ease of reference. Some valuers where large claims are involved, use quantity paper (which has several columns ideal for No. of worker days, tractor and implement days, spot items etc.). However, it is preferred that the claim should be kept as simple and clear as possible.

The item of dilapidation should always be first specified, followed by the remedy needed and the cost thereof e.g.

No. Ref	*Item*	*Rate*	*Total*
O.S. No. 6531—6.8 ha stubble			
1	Foul with Couch, Spray with Roundup 1990 6.5 ha		
1a	Spray with halfdose—do in 1991		
O.S. No. 6531/6888—Very high overgrown hedge			
2	Cut and lay 224 m		
O.S. No. 6531/6888 Wide blocked ditch			
3	Dig out and leave clean 200 m		

O.S. No. 7331—10.85 ha pasture

4 Pond in N.E. corner silted up. Clean out.
10 hrs drag line

11.5 COSTING

Costing claims accurately can be very difficult. There is such a variation in different parts of the country of relevant matters such as availability of labour, skill, types of soil, topography, location etc. The annual costings produced in July of each year by the Central Association of Agricultural Valuers is a most valuable guide and can be adapted for many of the items that are likely to arise in any claim e.g. if field is foul with couch the cost can be calculated on the basis of cost of spraying say £2.60 and cost of Roundup say £18.20 = total of £20.80 per acre or £51.40 per ha. If a further application at half rate is needed the following year the claim will be again £22.60 plus £9.10 for the spray—a total of £11.70 (£28.90 per ha).

11.6 Most of the work of remedying dilapidations is done by the farmer himself, his own staff and with his own equipment. Some items cannot be costed on unit basis and it is then normal to estimate the time to be taken on a worker day basis together with the use of any tractor and equipment needed e.g. to remove the concrete base of a former cow kennel building (a Tenant's fixture) sold off by the previous Tenant, but leaving the base would take, say 3 worker days for a farmer, together with 3 days' use of a tractor, loader, bucket and trailer. The claim would then be calculated as follows:

3 worker days @ £39.90	£119.40
3 days tractor @ £40	120.00
2 days F.E. loader and earth bucket @ say £15	30.00
1 day trailer @ say £15	15.00
	£284.40

11.7 HEDGES AND FENCES

Hedges become frequently overgrown, high, wide gappy and not stockproof. Often they are in need of cutting and laying. Some hedges are much easier to cut and lay than others e.g. hedges with a predominance of hazel and young thorn.

Others are very difficult to lay with many strong years growth that needs considerable axe work to lay and also cutting out with chainsaws. Note that where the tenancy agreement specifies cutting and laying a proportion of the hedges each year, this should be done. If there is no tenancy agreement, *S.I. 1473, 1973* Paragraph 9 operates, and this requires the cutting, trimming or laying of a proper proportion of the hedges in each year in order to maintain them *in a good and sound condition.*

Note, that the said Paragraph 9 specifies, as do many tenancy agreements, that *a proper proportion of hedges should be laid* annually. Others are trimmed and some left to grow for laying at a future date.

Occasionally a tenancy agreement my specify that a tenant shall plant up gaps that have developed, with quickthorn. Where this is not specified and gaps exist a claim will arise for gapping up with deadwood or possibly railing the gap or probably making the gap stockproof with posts and wire.

Woven wire fences deteriorate after a number of years (usually 15), the wire rusts and becomes loose and posts rot, frequently part or the whole, need replacing. In districts suffering from atmospheric pollution, or near the seaside, wire will rarely last 10 years.

Post and rail fencing again rots and becomes defective and needs replacement. Barbed wire is usually a Tenant's fixture but where this is established to be the Landlords' property, a claim will arise if posts rot and the wire becomes loose and rusts.

Hedges should be trimmed annually (where they are not left for growing on, to lay).

Frequently brambles and briars etc. encroach onto the adjacent field. A claim can arise for cutting these back.

11.7.1 *Specimen claims*

High very strong, wide and gappy hedge
Cost of cutting and laying .... m @ £

Overgrown hedge
Cost of cutting and laying .... m @ £

Hedge cut down with Shapeshaw but not stockproof
Cost of erecting alongside the hedge line a woven wire fence

on tanalised or pressure treated posts at 2m centres with single strand of barbed wire .... m @ £

Hedge has numerous gaps
Cost of planting two rows (staggered) quickthorn and protected on both sides with woven wire fencing with barbed wire strand .... m @ £

Pig wire fence rusted, weak and not stockproof
Cost of removing and replacing with post and medium gauge pig netting fence with two strands of barbed wire .... m @ £

Post, four rail fence defective in places and not stockproof
Cost of removal of defective sections and replacing with sawn tanalised replacement fence to match existing .... m @ £

Two strand barbed wire fence rusted and defective
Renew .... m @ £
Single strand renew .... m @ £

Hedge not trimmed
Trim and dispose of trimmings .... m @ £

Brambles, briars and other undergrowth encroaching at sides of hedge into adjoining fields.
Cost of cutting with flail cutter .... hrs @

11.8 GATES

Tenants are usually responsible for maintaining and replacing gates. Some old tenancy agreements still provide that the Landlord should provide materials for gates, in which case a claim would be restricted to the cost of making and erecting the gates. Some old gateways are still 10ft wide which is too small for modern equipment. The outgoing Tenant will not be responsible for providing a larger gate nor will he be responsible for providing better quality gates than previously existed e.g. a 15ft (4.55m) galvanised iron gate in place of a 10ft (3.03m) wooden gate. Gates should be complete with all hanging and securing fittings. Where gateways are depressed by stock traffic and a large gap exists under the hung gate, a claim can arise for building this up to its proper level with earth or stone as required.

Gates are frequently bent, broken, are not properly hung and drag on opening, and sometimes fittings are missing. Again, the hanging and shutting posts may need replacing and re-setting and side rails need repairing or replacing.

11.8.1 *Specimen claims*

3.66m (12ft) metal gate damaged beyond repair
Remove and replace with similar gate complete with fittings and properly hung.

Gate opening existing but 3.66m (12ft) gate and posts missing
Supply and erect 3.66m metal gate complete with hanging and shutting posts, together with all necessary fittings.

Top rail of gate and shutting post bent
Straighten out top rail and replace shutting post complete with catch.

Gate not hung properly
Take up loose hanging post, reset and rehang gate to close properly—

Securing catch to gate missing
Replace.

11.9 DITCHES, DRAINS AND CULVERTS

Ditches frequently become filled in, blocked, silted up, weeds grow on the sides and on ditch bottoms. Where not protected by barbed wire at the side, stock tread the sides in. Occasionally drain outfalls discharging into ditches also get trodden in and partly blocked. The remedy is to clean the same out and to protect (if stock is kept on the adjacent land) with a barbed wire fence. It should be noted that barbed wire fencing against ditches cannot be claimed for, unless it existed at the commencement of the tenancy. In times gone by, ditches were cleaned out by hand (and this may be necessary even today, where tractors cannot approach the ditch) but nowadays they are generally cleaned out, mechanically. The work is normally carried out by specialist contractors.

Culverts and bridges become damaged and occasionally need rebuilding and sometimes repairing.

Tenants are usually responsible for maintaining field drains and where the agreement is silent or does not exist, S.I. 1473 1973 provides that he shall be responsible for keeping them clear of obstructions. Blockages can sometimes be located and made good and the cost of making good is the amount of the claim. However, it occasionally will be cheaper and more effective to lay a new drain in a wet area than try to locate a blockage.

11.9.1 *Specimen claims*

Wide water, course (av. width $\frac{1}{2}$m) blocked and in need of cleaning
Cost of cleaning out deep wide watercourse and spreading soil
.... m @

Culvert (4 m) fallen in
Dig out and renew providing 3 No. .... mm dia. concrete pipes surrounded in consolidated shingle and finished with 6 in deep concrete surface

Ex-railway sleeper bridge (4 m) rotted and weak
Renew—2 w/d @
14 sleepers

Field drain blocked
Cost of unblocking—1 w/d

Wet area where drains not functioning
Cost of laying 80 mm plastic pipe drains 50 m @

11.10 FOUL LAND

Weed infestation can take several forms such as:

(a) Couch (or squitch) infestation in arable and pasture land
(b) Black grass in cereals
(c) Docks in pasture and arable land
(d) Wild oats in cereals
(e) Thistles in cereals and pastures
(f) Bracken in pastures
(g) Annual broadleaved weeds e.g. thistles, mayweed, poppy, charlock etc. in arable land
(h) Nettles.

(a) Couch

Most frequently this is found in cereal crops, although it infests any arable crops and is also found in pastures. The expensive additional cultivations and fallows of yesteryear are no longer necessary. Treatment is by spraying stubble with Roundup at 4 litres per ha. This costs approximately £51.50 per ha (£20.80 per acre), this being done when the weed is growing profusely after harvest. It is not necessary to leave the spraying until after harvest as pre-harvest spraying, at 2–4 litres per ha depending on the density of couch and can be done provided the crop is at the correct stage of maturity.

The use of additives to reduce the cost of Roundup application is commonplace today. However not all are approved by the manufacturers of Roundup.

If pastures are infested, complete kill of all couch and other grasses will be effected by application of Roundup at 4–6 litres per ha costing £51.50–74.40 ha (£20.80–£30 per acre).

(b) Blackgrass

Blackgrass infestation is an ever increasing problem in continuous cereal growing especially on heavier soils. It is controlled by spraying with Dicurane, either pre- or preferably post-weed emergence at a cost of approximately £33–£35 per ha (£13.50–£14.00 per acre). This spray, however, cannot be used on oats or certain varieties of wheat and barley. Arelan, Tolkan or Hytane can be used on all varieties of wheat or barley, either pre- or preferably post-emergence of weeds (post emergence application generally gives better weed control as with Dicurane) and the cost is £32.00 per ha.

(c) Docks

These are found in both arable crops and pasture. Control is by spraying with 2.8 litres per ha Asulox at a cost of £24.00 per ha. in pasture. In arable crops many herbicides will give partial control. Pre or post harvest use of roundup is a further method of control.

(d) Wild Oats

Wild oats have spread the breadth of the country, probably through purchased seed grain. To make a claim for wild oat infestation it will be necessary to inspect the crops on the farm concerned during the months of June to August, prior to the termination of the tenancy, when the infestation is apparent. However, care should be taken to find out whether the wild oats have been 'rogued' i.e. picked out by hand, prior to inspection. Wild oat seeds can remain dormant in the soil for years. Distinction should be made between wild oats and the occasional cultivated oat plants (from a previous crop) which are sometimes seen growing in a different crop and which may be odd seeds that have germinated. Wild oat plants have awns (or spikes) at the end of the seed whereas oats do not.* Wild oats can be controlled in cereals by spraying with Avenge 3.3–6.6 litre l/ha costing £29–52 l/ha applied. Commando @ 3 litres per ha costing £35 per ha applied (£14 per acre). Avadex can be used in a number of arable crops including cereals in either granule or liquid forms at varying rates costing £15 to £23 per ha applied (£5–£9.30 per acre). Hoegrass can also be used in arable crops @ 1.5–3.0 litre l/ha costing £25–45 l/ha applied.

(e) Thistles, Buttercup and General Pasture Weeds

These are treated by spraying with 2.8–3.5 litres M.C.P.A. per ha at a cost of £6.50–£8.00 per ha applied.

(f) Bracken

Where this is a problem, as is occasionally experienced in upland and rough pastures, it can be controlled by spraying with Asulox at a cost off £80 per ha (£32 per acre). It can be aerially sprayed but the cost will be much greater.

(g) Annual Broadleaved Weeds in Cereals

These include thistles, charlock, poppy, mayweed, chickweed, redshank, fat hen, corn marigold etc. Control is by spraying Panther/Pre-empt/Prebane (pre-emergent) at £24–31 per ha or Seloxone/Harrier/Crusader/Ally (all post emergent) at a cost of £18–31 per ha applied.

(h) Redshank

This weed can be a very serious problem in heavy soils in high rainfall areas. It is controlled in cereals by spraying, with the sprays

* Certain varieties of cultivated oat do have awns on some of the grains, but not all grains as with wild oats, also note hairiness of wild oat grains.

described in (g) or, if specific on its own, by spraying 2-4DP at approximately £9–£9.50 per ha applied (£3.80 per acre).

Many sprays used in root crops will control Redshank e.g. Semeron in kale at £28 per ha applied (£11.30 per acre).

(i) Nettles

These are often found in clumps in pastures, tend to spread and are difficult to control. Control is by spraying with Garlon costing £60 per ha applied (£24.25 per acre).

11.10.1 *Specimen claims*

Thistle and general weed infestation in pasture
Spray once with MCPA or MCPB @ ha

Couch infestation after winter wheat crop
Cost of spraying with Roundup @ £ per ha

Nettle infestation in pasture
Cost of spraying with

Dock infestation in pasture
Cost of spraying with

Annual weed infestation including thistle, charlock and mayweed in winter barley crop
Cost of spraying with

Severe wild oat infestation in all crops
Cost of two annual successive sprays with

11.11 PASTURE REPAIR

11.11.1 Often cattle are outwintered and cause damage to pasture and, in particular poaching occurs at gateways, feeding areas and also sometimes over the whole field through hoof marks, tractor and trailer wheel marks to cattle racks etc. Often the damage resultant is not sufficient to warrant ploughing up and re-seeding. The damage can usually be rectified by e.g. two disc harrows to the affected area, applying fertiliser, broadcasting grass seed, a light harrow and a rolling. Nevertheless, a loss will also result to the incomer in that this will take time to establish and he can suffer loss of grazing in the meantime. An example of the work necessary and its cost is:

	per Acre	*per ha*
2 disc harrows	6.50	16.00
1 fertiliser spinner	2.35	5.80
1 seed broadcast	2.20	5.45
1 light harrow	1.75	4.30
1 roll	4.20	10.35
Seed	20.00	50.00
Fertiliser 2 × 50 kg 20:10:10 @ £130	13.00	32.11
Electric fencing area off say	4.00	8.50
	£54.00	£132.51
But since this area is a small one, costs should be increased by 30% to say	£70.00	£173.00

If the damage is severe, it may be necessary to sub-soil the affected area.

Occasionally small rutted areas are found and these, as a rule, can be repaired by disc harrowing or rotovating, broadcasting seeds and rolling.

11.11.2 *Specimen claim*

2 ha pasture poached and to be re-established
Cost of cultivations, seeds, fertilisers and fencing off
2 ha @ £
Loss of grazing during re-establishment
4 weeks spring grazing @ £ per ha/per week

11.12 DILAPIDATIONS TO BUILDINGS ETC

11.12.1 A thorough knowledge of the Law of Dilapidations is necessary. As far as agricultural tenancies are concerned, the case of *Evans* v. *Jones* 1955 is relevant where the tenancy has only existed, for a short time. Here it was held that the Tenant could not have reasonably been expected, during the short term of the tenancy, to have to put the premises in repair, having regard to the condition at the outset and this must be taken into account when assessing the Landlord's entitlement to dilapidations where the tenancy is of a short duration. If the tenancy has existed a long period, this rule will be of no help to a Tenant.

11.12.2 In assessing building dilapidations, it is essential that all items should be measured up accurately and priced out in the proper way using an up to date builders' price book.

11.12.3 The claim will be assessed under the terms of the tenancy agreement or under S.I. 1473 1973 dependent on whether the claim is under S.71 (1) or S.71 (3) of the 1986 Act. If under S.I. 1473, its provisions must be studied very carefully and in particular paragraph 11, which clearly states that any accrued liability on the part of the tenant in respect of internal and external decorations can be the subject of claim.

11.12.4 Assuming repairing liabilities are in accordance with S.I. 1473 1973 the tenant is responsible for a large number of repairs including inter-alia, boilers, fireplaces, drains, manholes, water supply systems (above ground), tanks, troughs, pumping equipment, cattle grids, bridges, ponds, roads and yards, He is also to keep clean and in working order roof valleys, gutters and r.w.d.p. wells, septic tanks and cesspools and also to pay half the cost of repairs to doors, windows, interior staircases, eaves guttering, downpipes, floorboards and skylights. Note, however, that by virtue of 1 (3) Part I of the S.I., the tenant is not responsible for replacing items mentioned in 5 (1) Part II e.g. boilers, ranges, water supply systems, tanks, pipes etc. which have worn out or become incapable of further repair, unless the tenant is responsible for its replacement under Para 6 (where the replacement is necessary due to the wilful neglect of the tenant etc.). Reference is also made to the case of *Robertson Aikerman* v. *George* (1953) decided by Leicester County Court that the Landlord could not recover from the tenant the one half cost of repairs of floorboards, doors, etc. unless the work had been actually done at the termination of the tenancy. It appears, therefore, that the Landlord should ensure that work, where he would normally recover half cost, should be undertaken *prior to the termination of the tenancy*.

11.12.5 *Specimen claims*

Central heating boiler not functioning
Repair and put in working order

Ball valve to automatic water trough in yard leaking
Repair and put in working order

5 No. tubular bars to cattle grid at farm entrance bent
Take out and renew

Right hand stone parapet to bridge on farm drive damaged

Take down and re-build

All eaves and gutters on south elevation of farmhouse blocked
Clean out gutters, and leave in working order

Septic tank to farmhouse overflowing
Clean out and leave in working order

Farm drive severely potholed
Cut out potholes and fill with tarmacadam, properly consolidated to level of surrounding road surface

Lounge in need of redecoration. Redecorated over 7 years ago
Redecorate including all preparation work, stripping wallpaper, stopping and making good, rubbing down, preparing and applying 2 coats oils to all previously painted wood and iron work, twice whitening ceilings and re-papering walls—Say (If redecorated say 4 years ago the claim would be £ $\times \frac{4}{7}$ = £

No. 2 window panes cracked in bedroom No. 1
Hack out, prepare rebates and reglaze in 4 mm sheet glass—0.4 m^2 @ £

No. $1\frac{1}{2} m^2$ of boarded floor to Granary rotted due to tenant keeping poultry in the building)
Take up defective area and renew m @

11.13. OTHER CLAIMS

Other claims may arise contractually, and under Section 11 of the Agriculture Act 1947. These may embrace.

11.13.1 Shortage of New Seeds

This claim may arise both under Section 11 and also contractual obligation. If a Landlord's interest is damaged through such seeds not being left, the claim will be for making good such shortage. However, in computing such claim, consideration should be taken of the possibility of land growing such seeds being available for growing other crops.

11.13.2 Departure from Rotation

The Tenant has generally a freedom to crop arable land as he wishes, except in the last year of the tenancy. However, if this system of cropping causes injury or deterioration to the holding, the Landlord can recover damages under Section 11 of the 1947 Act and also both Sections 71 and 72 of the 1986 Act.

11.13.3 Loss of Quota

No claim can arise here except where there is a specific breach of the Contract of Tenancy or loss of quotas which the Tenant cannot dispose of without his Landlord's consent.

11.13.4 Selling Produce in the Last Year of Tenancy

The Tenant is not allowed, by virtue of Section 11 of the 1947 Act, to sell off or remove produce in the last year of tenancy. If so removed, the Landlord can claim for the equivalent manurial value of the crops sold off. This applies only to produce normally consumed on the holding such as fodder but not corn.

11.13.5 Continuous Cereal Growing

Damage can be caused to the freehold by continuous cereal growing resulting in infestation of wild oats, couch, blackgrass, pests, loss of fertility, disease etc. The soil structure may also be damaged. Claims may arise under both section 71 and 72 of the 1986 Act.

11.13.6 General

The Landlord's claim in respect of the above and possibly other matters may not be properly compensated under Section 71, in which case claims should also be made under Section 72 for deterioration. However, for such a claim to succeed it is reminded here that a proper notice of intention of claiming under this Section should be served before the expiration of one month *before* the tenancy terminates. The compensation recoverable for a deterioration claim is restricted by virtue of Section 72(3) to the decrease in the value of the holding, as a holding having regard to the character and situation of the holding and the average requirements of tenants skilled in husbandry.

Example of Dilapidations Claim

AGRICULTURAL HOLDINGS ACT 1986

Re: The Holding known as:

WEST FARM, METCHINGFIELD, HEREFORD

To: Mr. J. Yeoman-Farmer,
Holly View,
Dinedor,
Hereford

(Outgoing Tenant)

WE HEREBY GIVE YOU NOTICE pursuant to Section 83 (2) of the above Act of our intention to make against you certain claims arising out of the termination of the tenancy of the above holding the nature of which claims is set out in the Schedule hereto.

THE SCHEDULE

DILAPIDATIONS:
Claims made under clauses 111 (4) (9) (10) (19) (27) of the Tenancy Agreement dated 4th April, 1972 and also under S.I. 1973 No 1473

FARMHOUSE

1. All eaves gutters require cleaning out. Clean out. £15.00
2. Septic tank blocked up. Clean out £30.00
3. Bedrooms No 5 (NE) and 6 (SE) require redecoration. To cost of stripping walls, making good, preparing and painting all wood and iron work previously painted two coats oil and re-papering walls with comparable quality paper. £600.00
4. W.C. pan cracked in Bathroom. Replace £50.00

5. Hot water tap in Cloakroom leaking. Replace washer £5.00
6. Garden, Vegetable garden very unkept and weed infested. Dig, leave tidy and control weeds. £50.00

FARM BUILDINGS

7. Dairy	No 5 window panes broken. Renew	£35.00
8. Range of 3 Calving Boxes	(a) No 2 water bowls missing. Replace and re-connect supply	£40.00
	(b) Half heckdoor to middle Box damaged beyond repair. Cost of replacement inc. fittings	£40.00
	(c) Gulley grid missing. Replace	£10.00
	(d) Manhole cover cracked. Replace	£50.00
9. Range of No 4 Calf Cots	Limewash to walls to renew. Cost of cleaning and carrying out necessary work. 2 W/D and Materials	£90.00
10. Barn	Renew missing and loose slates Cost of renewal (Maximum permitted)	£100.00
11. Pump House	'Hydra' water pump not working. Cost of putting into working order. (estimate attached)	£126.50
12. Slurry Lagoon	Full up. Cost of cleaning out. 4 W/D (with agitator, tractor, pump and slurry tank and spreading) @ £95	£380.00
13. Electrical	No 1 switch in Piggery broken and No 3 electric pendants in covered yard missing. Cost of putting into repair.	£48.20
14. Farm Drive	Severely potholed in many places. Cost of cutting out and filling with tarmac and rolling.	£250.00

LAND

Pt. O.S. No 602 8.79 ha Winter Barley

15. 602/S.E.	High overgrown hedge. Cost of cutting and laying 323 m @ £2.50	£807.50
16. 602/S.E.	Wide ditch blocked. Cost of cleaning out 310m @ £1	£310.00
17. 602/652	Culvert broken. Cost of reconstructing.	£150.00

Pt. O.S. No 652 3.32 ha Winter Barley

18.	Severe infestation of wild oats. Cost of spraying with Avenge in 1990 and 1991. $3\frac{1}{4}$ ha @ £80.	£260.00
19.	Land drain blocked in S.E. Corner. Cost of locating blockage and making functional.	£50.00

20. 652/220	108 m run of gaps in hedge. Cost of making stock proof @ £2	£216.00
21. 652/220	Top bar of metal gate bent. Cost of making good.	£10.00

O.S. No 220 7.42 ha Winter Wheat

22.	Severe infestation of couch. Cost of spraying with 'Round up' after harvest of 1990 and cost of ditto spraying with half strength dose in 1991. 7¼ ha @ £77	£558.75
23. 220/218	Post and rail fence rotted. Cost of removal and renewal 156 m @ £6.50	£1,014.00

O.S. No 218 6.87 ha Pasture

24.	Infestation of docks to approx 2 ha in N.W. corner. Cost of spraying twice @ £48	£96.00
25. 218/618	Hedge not trimmed and not left for growing to lay. Cost of trimming 246 m @ 7p	£17.22
26. 218/216	No 2 Hedgerow Oak Trees with barbed wire nailed thereto. Loss of timber.	£50.00

O.S. No 618 4.56 Roots and Stubble

27.	Uncultivated area of ¼ ha in S.W. corner (site of dung heap) full of weeds. Cost of eradicating weeds, cultivating and bringing back into proper production.	£50.00
28. 618/653	Dry stone wall fallen into disrepair. Cost of rebuilding 10 sq m @ £20	£200.00
29. 618/653	Gate opening where gate previously existed but no gate or posts. Cost of replacing 3.6 m metal gate and posts all complete	£120.00

O.S. No 832 5.25 ha Ploughed

30. 832/618	Overgrown hedge. Cost of cutting and laying 289 m @ £2.30	£664.70
31. 832/618	Area of briars encroaching into field at headland at S. end. Cost of cutting back and burning. 1 W/D	£39.80
32. 832/618	Shutting post missing to gate Cost of replacing	£25.00
33.	Pond in S.E. corner silted up and outlet drain blocked. Cost of cleaning out and unblocking drain. 8½ hours with mechanical digger and 2 hours tractor and trailer	£150.00

O.S. No 54 9.10 ha Pasture

34.	Whole field infested with thistles. Cost of spraying @ £7	£63.70

35. 54/53	Metal gate damaged beyond repair by vehicle collison. Cost of replacing	£50.00
36. 54/53	Side rails both sides gate opening disrepair. Cost of removing and replacing 6 m @ £6	£36.00
37. 54/53	Small ditch silted up. Cost of cleaning out and spreading spoil 171 m @ 50p	£85.50
38. 54/53	Landlords barbed wire fence alongside ditch in disrepair. Cost of re-erecting and replacing No 25 fencing stakes	£35.00

O.S. No 651 7.03 ha Pasture

39.	Approximately 4 ha of Pasture infested with clumps of nettles. Cost of spraying and restoring infested areas to proper production. 4 ha @ £60	£240.00
40.	Whole field infested with thistles. Cost of spraying @ £7	£49.21
41. 651/654	Lane on boundary to part overgrown. Cost of clearing and controlling weeds	£100.00

O.S. No 654 5.68 ha Pasture

42.	Approximately 1½ ha infested with docks. Cost of spraying with Asulox three times to eradicate and loss of production. 1½ ha @ £72	£108.00
43.	Approximately 1¾ ha infested with couch on N. boundary alongside hedge. Cost of spraying	£90.00
44.	Approximately 1 ha severely poached by outwintered cattle. Cost of restoration and loss of production	£200.00

RESERVATION. The right is reserved to amend the above claim at any time prior to settlement or arbitration.

Dated this fifth day of February 1990.

....

WILLIAMS, DAVIES & REES, Agricultural Valuers
DOLANOG,
WELSHPOOL,
POWYS.

Duly Authorised Agents and Valuers on behalf of the Walford Estate Company. Landlords.

E. & O.E.

PRACTISE NOTE: ITEM 25: Since this is the only hedge to trim and having regard to the extra costs of getting a contractor onto the farm to do this work, the normal charge is doubled.

CHAPTER TWELVE

Pipelines, Capital, Easement Payments & Claims

12.1 GENERAL

Claims arise following compulsory rights orders to lay sewers and mains water under the Water Act 1989 Schedule 19(4), gas mains (under the Gas Act 1986—Section 70 & Schedule 3), commercial pipelines (e.g. oil) under the Pipelines Act 1962. Three types of claim arise:

(a) easement claims for gas and commercial pipelines
(b) capital payments for sewer and water mains (recognition payments)
(c) Damage claims arising following the laying of any pipelines

12.2 Capital Payments

These arise following the laying of sewer and water mains and are a 'recognition payment' for the presence of the pipeline as no easement is taken. In the past it has been the practise of the authorities to pay 50% of the capital value of the freehold strip affected. Water Authorities usually contended that this should be restricted to narrow notional easement widths occupied by the pipes usually from 3½m to 5m. However the recent case of *St. Johns College Oxford* v. *Thames Water Authority* (Lands Tribunal 1990) decided that payment was to cover 50% of the full working width of 20 yards and held that the damage on injurious affection caused by the pipe affects not mainly the strip but each farm as a whole.

Additional payments are made for marker points (usually £10 each) and also for chambers, valves, manholes and other permanent structures. If these manholes etc. are in arable fields they are a serious nuisance and it is not unusual especially in the case of the larger chambers for compensation from £100 to £500 each to be paid for their presence. If the chambers are placed in hedgerows the interference is less as is the compensation e.g. £30 each. Sometimes an authority can be persuaded to bury a manhole below the depth of cultivation and this will result in lesser compensation payable.

12.3 Easement Payment

These payments arise when commercial pipelines are laid e.g. oil and for gas mains. British Gas and promoters of commercial pipelines normally pay an easement payment of 75% of the vacant possession value of the strip of land affected. Quite often, dependent on the width of easement these have been up to £10 per yard run.

In addition an 'occupiers payment' is made to the occupier (whether the owner or tenant) hopefully to obtain co-operation. This payment is normally agreed between the promoters and the CLA & N.F.U.

Again payments are made for manholes, chambers and other structures.

12.3 Damage Claims

These should cover the following items:

- (i) Actual crop losses
- (ii) Future crop losses—over a number of years from, say 2 years to 10 years, dependent on damage caused when the work was done.
- (iii) Cost of restoration
- (iv) Cost of replacing lost fertility
- (v) Weed control. Weeds have a habit of spreading on pipe-tracks and also being introduced to clean land where they did not previously exist
- (vi) Cost of future hedge-laying costs where hedges are re-planted. This is usually in 10/12 year's time
- (vii) Cost of replacing top soil where subsidence occurs after a period of time. Also for replacing any top soil lost
- (viii) Drainage. Frequently drains are cut. It is important that they should be properly reinstated after the main is laid. In small diameter pipelines, a generally satisfactory method is for a section of plastic pipe to be laid on a substantial 4″ × 4″ hardwood baulk which is let into the soil for ½m either side of the trench, and connected to the exposed drains. In the case of large diameter pipes, some support

such as concrete built up from the main to support the actual replacement drain, is needed. In any event, the authority laying the main should be held responsible at all times for any damage caused to drains as a result of their operations.

(ix) The claimant should be paid for all the time he spends on the matter e.g. with agents, contractors, valuers, solicitors, rounding up stray stock, telephone calls, travelling expenses etc. A detailed diary of time used should be kept.

(x) All professional fees

(xi) Interest on the agreed claim. However, it is usually difficult to recover this from the pipe-laying authority, except on the easement payment.

12.4 Example Easement Recognition Payment

283 m run of 150 mm dia. water mains is laid through a farm. The working width is 15 m. There are four marker points and three valve chambers (one in a hedgerow). The suggested claim is:

(*a*) *Pipeline*
283 m × 15 m = 4245 sq.m. or 0.424 ha.
0.424 h @ £6,175 (£2,500 per acre) × 50% = £1,309.10

(*b*) *Marker Points*
4 No @ £10 = £40.00

(*c*) *Valve Chambers*
3 No @ £75 £225
1 No @ £50 50 = £275.00

(d) *Fees*
Valuers & Solicitors Fees

(e) *Interest*—on the claim from date of entry to date of payments.

Example of Pipeline Claim

STATEMENT OF CLAIM

by

MR. R. J. BENGOUCH

against

THE VALE OIL CORPORATION LIMITED

Following the laying of an Oil Pipeline through WEST FARM, METCHINGFIELD, HEREFORD

March 1990 PREPARED BY: WILLIAMS DAVIES & REES,
AGRICULTURAL VALUERS,
DOLANOG,
WELSHPOOL,
POWYS

1. LOSS OF CROPS

1.1) O.S. 1481—Working area 0.24 ha

a)	1989—Loss of Copain Winter Wheat 7.4 tonnes per ha @ £120 per tonne	£ 213.12
b)	1989—Loss of straw 3 tonnes per ha @ £25 per tonne	18.00

1.2) O.S. 8352—Working Area 0.30 ha

a)	1989—Loss of Desiree Potato Crop 50 tonnes per ha @ £130 per tonne	1,950.00
b)	1988—Loss of Triumph Spring Barley 4.9 tonnes per ha @ £120 per tonne	176.40
c)	1988—Loss of Barley Straw 2.5 tonne per ha @ £25 per tonne	18.75

1.3) O.S. 1545—Working Area 0.4 ha

a)	1988—Loss of Sugar Beet Crop Site works made it impossible to harvest the sugar beet in the area of the field south of the easement Total area lost 1.58 ha 54 Tonnes per ha @ £30 per tonne (net of transport)	2,559.60
b)	1988—Loss of Sugar Beet tops 1.58 ha @ £75 per ha	118.50
c)	1989—Loss of Triumph Spring Barley 5 tonnes per ha @ £120 per tonne	240.00

1.4) O.S. 2400—Working Area 0.42 ha

a)	1988—Loss of Winter Grazing @ £99 per ha	41.58
b)	1989—Loss of Spring, Summer and Autumn grazing @ £346 per ha	145.32

1.5) O.S. 6633

1988—Loss of Avalon Winter Wheat. Site works made

it impossible to harvest area of crop to the south of the easement. Total area lost 0.121 ha. Loss of 10 tonnes per ha @ £120 per tonne. 145.20
Loss of straw 8.00

1.6) O.S. 9500

1988—Works to the Main caused a loss of .2 ha of Sugar Beet in the corner adjoining O.S. 1545 and O.S. 8352.
Loss of 54 tonnes per ha @ £26.73 per tonne 288.64
Loss of Beet Tops @ £75 per ha 15.00

2. REINSTATEMENT OF HEDGES

O.S. 8352/9500	18 m
O.S. 9500/1545	15 m
O.S. 1545/2400	30 m
O.S. 2400/6633	18 m
O.S. 6633/ ×	18 m
O.S. 4500/Railway	16 m
	115 m

To cost of first laying of re-planted hedges in 10 years. 115m @ £8 (allowing for inflation and deferring) 920.00

3. REINSTATEMENT OF WORKING AREA

3.1) O.S. 4500, 8352, 1545 and O.S. 633: TOTAL AREA 1.01 ha

To cost of additional expenses in cultivating and working down the area affected by the laying of the main 1.01 ha @ £86 per ha 86.86

3.2) O.S. 2400—Area 0.42 ha

To cost of reinstating the working area including the levelling of the surface, panbusting, cultivating, and re-seeding (two sections) 0.42 ha @ £321 per ha 134.82

3.3) Restoration of Fertility

The Corporation to pay for the cost of restoring fertility to the areas affected by the laying of its main. This involves the application of compound fertiliser, lime and farmyard manure. The claimant has had the trench area tested for acidity and a PH value of 4.8 has resulted. It is recommended that $12\frac{1}{3}$ tonnes of lime per ha be applied in two doses over two years.

$12\frac{1}{3}$ tonnes lime @ £13 per tonne	160.29	
25 tonnes F.Y.M. @ £12 per tonne	300.00	
12 × 50 kg 15–15–15 Compound	84.00	
1.43 ha @	544.29	778.33

3.4) Removal of Surface Debris

The Corporation to pay for the cost of removal of rock and other debris brought to the surface following the laying of the main.
3 worker days plus tractor and trailer @ £95 per day — 285.00

3.5) Removal of Wire Fencing

The Corporation to pay for the cost of removing the wire fencing marking the boundary of the working area
2 worker days plus tractor and trailer @ £95 per day — 190.00

3.6) Weed Control

The Corporation to pay for spraying the affected area with a herbicide to control the inevitable growth of weeds
1.43 ha @ £74 per ha — 105.82

4. REINSTATEMENT OF DISUSED RAILWAY LINE

4.1) Repairs—The old railway line has been damaged by the contractors and is now very rutted and holding surface water. The bank against O.S. No 4500 is badly reinstated and a large heap of earth left.
To cost of reinstatement including levelling and stoning — 400.00

4.2) Fence

A Double set of post and rail fencing should be provided between the old railway line and O.S. 4500 and planted with a quick thorn hedge, or alternatively the Corporation to pay compensation
15 m @ £15 per m — 225.00

5. CONSEQUENTIAL LOSSES

5.1) Sheep Damage

a) Sheep got out of O.S. 2400 and O.S. 1545 on three occasions due to the inadequacy of the fence to the working area. They entered O.S. 9500 and ate the emerging Sugar Beet Plants.
Losses on 4 ha of Sugar Beet estimated at 30 tonnes @ £30.00 — 900.00
Loss of Sugar Beet tops estimated — 100.00

b) The sheep also disturbed the pre emergent spray in O.S. 9500 and it did not work, resulting in a weed problem
To cost of 2 steerage hoes on 8 ha @ £15 per ha

c) The sheep grazed the pasture in O.S. 6235 in the winter, but due to the inadequacy of the fencing to the working area they escaped into O.S. 4500 where they grazed some 8 acres of Avalon winter wheat.
Loss of 2 tonnes @ £120 per tonne — 240.00

d) One 3 year old mule ewe fell into the trench and died
To loss of ewe £75.00
Loss of lambs
180% lambing percentage
@ £42 per lamb £75.60 150.60

5.2) Damage to Roadway

The Contractors used the farm track in the winter and damaged the same.
To cost of making good track including labour £600.00
To charge for allowing contractors the use of the track £300.00 900.00

5.3 Disturbance to Rotation

a) The farm rotation has been disturbed in O.S. 8352, 8358/7832, making a total of 6 ha. Due to this Triumph spring barley had to be planted instead of winter wheat. A loss occurred on this disturbance to the rotation. Spring barley yielded 5 tonnes per ha as against a wheat yield of 7.4 tonnes per ha
Loss of 2.4 tonnes per ha @ £120 per tonne 1,728.00

b) The trench excavation prevented flood water draining away in O.S. No 4500—Copain Winter wheat. As a result 1.6 ha were lost and has to be re-drilled with spring barley.
Loss of 4 tonnes @ £120 per tonne plus cost of resowing 560.00

6. DISTURBANCE LOSSES

6.1) O.S. 1545—Sugar Beet

Due to laying of main it was additionally expensive to harvest (owing to extra time turning of harvester, etc.) the sugar beet in OS 1545
To additional cost 200.00

6.2) Disturbance

The Corporation to recompense the claimants for disturbance and inconvenience suffered resulting from the works.
The sum of 600.00

7. LOSS OF FUTURE YIELDS & PROFITS

Due to the very difficult working conditions arising from the wet autumn of 1988 and the operations generally in 1989 serious damage was caused to the soil structure as a result that the restoration works proposed will not fully indemnify the claimant from future losses of yield and consequently profits viz.

1st year 60% of £750 per ha.	£450	
2nd year 30% of £750 per ha.	£225	
3rd year 10% of £750 per ha.	£75	
1.43 ha @	750	1,072.50

8. EASEMENT PAYMENT

The Corporation to pay the claimant compensation as follows:

8.1) Pipeline Easement (Freeholder)

931 m × 20 m = 1.862 ha	
1.862 @ £6,175 (£2,500 per acre) × 75% =	8,623.38

8.2) Marker Points

6 no @ £10	60.00

8.3) Chambers

No 3 Chambers @ £100	
No 1 Chamber @ £20	320.00

8.4) Occupiers Payment

at the scale agreed with the N.F.U.

9. ADDITIONAL ITEMS

9.1) Drainage

The Corporation to make good at all time damage caused by the operations to the land drainage system, whether artificial or natural, and also to pay for losses caused by such damage and also by the restoration work put in hand.

9.2) Subsidence

The Corporation to make good any subsidence resulting from the mains laying operation, including provision of a minimum of 0.3m depth of good quality top soil of a quality equivalent to that lost where necessary.

9.3) Fees

The Corporation to pay the claimant's Valuers fees and expenses on the approved scale, together with Solicitor's costs.

9.4) Interest

The Corporation to pay interest at the statutory rate on the agreed claim from the date of entry.

9.5) Any Other Matters

E & O.E.

Practice Note

As a rule the Easement Payment and payment for Chambers and the Occupiers payment is agreed before entry on to the land. Marker point payments are usually dealt with at the time of surface damage claim is submitted. However, they are shown here in this specimen claim to illustrate the type of claim made.

CHAPTER THIRTEEN

Electricity Line Wayleaves and Claims

13.1 GENERAL

Important compensation claims usually arise on the construction of 400kv and 132kv overhead transmission lines. The former are usually erected by the Central Electricity Generating Board and the latter erected or maintained by area Electricity Boards, who are also responsible for smaller transmission lines. Claims take two forms:

13.2 LOSSES ARISING AS A RESULT OF CONSTRUCTIONAL WORK

This claim covers crop losses, future crop losses and all other losses arising on the line being erected. This also covers timber cut down, trees lopped, damage to soil structure (frequently this arises as a result of heavy equipment severely rutting and consolidating the area concerned). A claim will broadly follow the specimen claim given in respect of the laying of a main.

13.3 DAMAGE TO THE FREEHOLD

Electricity lines are unsightly, mar the landscape, interfere with views from farmhouses and cottages, interfere with irrigation systems, sporting and generally depreciate the value of a freehold. These items cannot, as a rule, be claimed for unless the claimant enters into a Deed of Grant, granting a perpetual easement for the line in return for compensation, The claim is usually assessed as follows:

(i) The annual rental payment is capitalised by multiplying × 20 acres purchase. (Where there is a second line in the same field of 20 acres or less, the capitalisation is increased by 50%). This is usually the *arable* rate of payment unless it is clear that the land concerned is not arable land and cannot be ploughed e.g. Hill Land.

(ii) Dwellings are valued and if the line is near and the view marred a percentage is agreed as being the depreciation in value e.g. a farmhouse may be worth £150,000 and the depreciation is agreed at say 5% or 10% or more. Cottages may be assessed in a similar way. The practical difficulty is generally in agreeing the percentage depreciation.

(iii) Sporting rights may be seriously interfered with e.g. a line going through an excellent shoot or over some good fishing pools on a river. This should be assessed as a reduction in rental, which reduction is capitalised at 20–25 Y.P.

(iv) Irrigation cannot be practised under overhead transmission lines for obvious reasons. Not every farmer irrigates nor has the need to, but if he does the interference and loss can be considerable, as irrigation is normally only used for high value cash crops such as potatoes and market garden crops. An estimate of the average annual net loss should be made, over the area affected, and this capitalised by such a Y.P. as would be used for valuing a farm, say, 25 to 30. However, this is a most contentious and arguable claim.

(v) Where a line passes through a wood, all timber is cut and the area affected is perpetually sterilised i.e. trees cannot be grown on the strip concerned, which can be as much as 50–60 ft wide. Total losses for the future arising from this should be claimed i.e. the profit on a crop of trees and also the danger to the surrounding woodland from windblow as the canopy has been disturbed.

(vi) Any other matters affecting the freehold e.g. buildings cannot be erected under overhead lines—loss of development values etc. etc.

13.4 Reference is made to the case *George* v. *S.W. Electricity Board* and *George* v. *C.E.G.B.* (E.G. page 355 Vol. 264, 23rd October 1982) when the Lands Tribunal decided that a claim could be on the 'elemental approach' (as above) or on the 'overall view' i.e. taking a percentage depreciation of the agreed value of the holding, and in this case awarded 5% of the value of the holding of £176,600. The two methods should be used to check each other.

13.5 The decision in the 1983 Land Tribunal case of *Clouds Estate Trustees* v. *S. Electricity Board* (E.G. page 367 Vol. 268) should also be considered. It emerged here that the annual payments or rents agreed between the electricity authorities and the N.F.U. and C.L.A. included nothing in respect of loss of visual amenity, nuisance value, damage to shooting etc. and these, and possibly other relevant matters, could be the subject of an additional claim. Care should therefore be taken that these matters are not overlooked when negotiating a permanent easement claim.

13.6 In every case, solicitor's and valuer's fees, interest at the statutory rate should be claimed, and also, in the case of the Damage Claim, payment for the claimant's time (which is usually considerable), spent in dealing with the matter.

CHAPTER FOURTEEN

Notices to Remedy Breaches of Tenancy Agreement

14.1 The Law on this subject is contained in the Agricultural Holdings (Arbitration on Notices) Order 1987 (S.I. 1987 No. 710) and Schedule 3, Part I Case D of the Agricultural Holdings Act 1986.

14.2 The service of an appropriate notice to remedy breaches is a method whereby a landlord who considers his tenant is in breach of the terms of his tenancy can possibly get the breach remedied. However, service of notices of this kind often gives rise to much contention and costs, and should not be embarked upon unless there are good grounds of serving such notice. A very careful study of the law is required to avoid pitfalls, many of which exist. Photographs of some of the defects taken at the time of preparation of the Notice can, in subsequent events, be of great help.

14.3 The notice served must be in the prescribed forms. There are two officially prescribed forms to use under S.I. 1987 No. 711 The Agricultural Holdings (Forms of Notice to Pay Rent or to Remedy) Regulations 1987 viz:

(a) *Form 2*—a notice to be served where the tenant is in breach of agreement and is required to do work of repair, maintenance and replacement
(b) *Form 3*—a notice where a tenant is required to remedy a breach of agreement and is *not* a notice requiring work of repair, maintenance or replacement e.g. to reside in the farmhouse.

14.4 In serving any notice, the prescribed notes must be attached to the notice. See notes under the example Notice (2) and also the different Notice (and note beneath) in the case of Form 3.

Where the notice is to do work, a period of less than six months shall not be treated as a 'reasonable' period.

14.5 In a notice to do work, it is important to clearly state, separately, the breaches and also, following on, separately again, the remedy required. The remedy required should be fair, reasonable and practical as, otherwise, an arbitrator can be expected, on a reference, to adopt a fair and practical attitude in his award. It would be unreasonable, for instance, to expect a tenant to lay hedges in the summer months. It is always wise to attach a plan to any notice to do work, served, and this should be numbered to correspond with the numbers given on the Notice.

It is also prudent to state that it is expected that the necessary work shall be executed to a proper standard and in a workmanlike manner.

14.6 *Example Landlords Notice* to Remedy Breach by doing work of repair, maintenance or replacement.

AGRICULTURAL HOLDINGS ACT 1986

Schedule 3, Part 1, Case D.
Notice to tenant to remedy breach of tenancy agreement by doing work of repair, maintenance or replacement.

Re: The holding known as
PART OF LOWER BARN FARM,
GREAT ONEN,
MONMOUTH,
GWENT.
O.S. NOS. 231, 231 & 228

To: MR. ALBERT CHARLES
GREAT ONEN FARM,
MONMOUTH,
GWENT.

1. I hereby give notice that I require to remedy within twelve months, from the date of service of this Notice the breaches set out below, of the terms or conditions of your tenancy, being breaches which are capable of being remedied of terms or conditions which are not inconsistent with the fulfilment of your responsibilities to farm the holding in accordance with the rules of good husbandry.

2. This Notice requires the doing of the work of repair, maintenance or replacement specified below.

PARTICULARS OF BREACHES OF TERMS OR CONDITIONS OF TENANCY

Terms or Condition of Tenancy Clauses 5(7), 5(8), 5(9), 5(14) etc. of the Tenancy Agreement dated 31st December, 1952. Also the rules of Good Husbandry as set out in section 11 of the Agriculture Act, 1947 and Statutory Instrument, No. 1473. The Agricultural (Maintenance, Repair and Insurance of Fixed Equipment) Regulations 1973.

Particulars of Breaches and work required to Remedy them

No. on attached plan	O.S. No. of field	
1	231	Pasture Foul—Eradicate couch, argostis, docks, thistles, chickweed etc., to bring field back into full production
2	231/243	Nettles, bracken and briars encroaching into field. Eradicate and control to bring area affected back into proper production.
3	231/243	Weak gappy hedge. Cut and lay and render stockproof
4	231/199	Ditch overgrown and not cleaned out. Remove overgrowth, shrubs and briars and clean out to proper fall
5	231/199	Weak gappy overgrown hedge. Cut and lay approx 17 yards of hedge, provide and erect a post and wire fence to the remainder to render the boundary stockproof
6	231/199	Bracken and briars encroaching into field. Eradicate and control to bring back into proper production
7	231/199 West end	Extensive growth of scrub and saplings. Cut and remove and provide proper fence
8	231/Roadway	Weak, gappy, overgrown hedge. Cut and lay approx 50 yards of hedge, and provide post and wire stockproof fence to the remainder
9	231/Roadway	Ditch overgrown and not cleaned out. Dig and clean out ditch to proper falls

10	231/Roadway	Undergrowth, bracken and briars encroaching into field. Eradicate and control to bring back into proper production
11	231/Roadway South of Lane	Weak, gappy hedge. Cut and lay and provide approx 43 yards of post and wire stockproof fencing
12	231/Roadway	Undergrowth and bracken encroaching into field. Eradicate and control and bring back affected area into proper production
13	231/Roadway	Barbed wire attached to Oak tree. Remove
14	231/232	Encroachment of undergrowth, brambles, weeds and thistles into the field and at the boundary. Eradicate and control and bring back into proper production
15	231/232	Ditch overgrown and not cleaned out. Dig and clean out ditch to proper falls
16	232	This Pasture field is currently in Arable Production contrary to its scheduling on the tenancy agreement. Return to pasture and in doing so eradicate couch, argostis, buttercups, mayweed and also other arable weeds present
17	232/Roadway West	Nettles, briars, undergrowth encroaching into the headland of this field. Eradicate and bring back into proper production
18	232/Roadway West	Ditch overgrown and hedge weak and gappy. Dig out ditch and fence hedge line with post and wire stockproof fence
19	232/228	Ditch and surrounding area overgrown with undergrowth, briars, bracken and thistles. Dig out ditch, spread soil, cut burn and eradicate undergrowth and bring whole area back into proper production

20	228	Stubble foul. Eradicate and control couch, argostis, docks and buttercups and other weeds
21	228/Lane	Encroachment of hedge and scrub woodlands into field. Cut and burn some 20 yard widths of thorn bushes, saplings, (mainly ash) dead wood, brambles, bracken, thistles, and nettles, eradicate and control to bring area back into proper production
22	228/Lane	Ditch overgrown. Dig and clean out to proper falls
23	228/227	Undergrowth, brambles, nettles, docks (approx 20 yards into field) encroaching into field. Cut, burn and control to bring back into proper production
24	228/234	High weak gappy hedge with blocked ditch either side. Cut and lay hedge making good the gappy area with stockproof post and wire fences, dig and clean ditches to proper falls
25	228/231	Undergrowth, briars, docks, nettles and fallen trees encroaching onto field, with weak unlayable hedge behind. Cut, burn and eradicate and control and bring area back into proper production. Provide post and wire stockproof fence on hedge line

3. This Notice given in accordance with Case D in Part I of Schedule 3 to the Agricultural Holdings Act 1986 and failure to comply with it within the period specified above may be relied on as a reason for a notice to quit under Case D.

4. Your attention is drawn to the Notes following the signature to this Notice.

RIDER

1. The above work to be carried out in a proper workmanlike and satisfactory manner.
2. Hedges are to be laid in the proper season.
3. Where ditches are cleaned out these must be cleaned to proper depths & falls & spoil to be properly spread.

4. Where encroaching undergrowth to be cleared and eradicated all roots to be removed and area levelled as necessary.
5. Where hedges have been neglected and it is impossible to find suitable materials to lay the same, post and wire fencing (pig or sheep netting and one strand of barbed wire fixed on chestnut posts or tanalised posts) at 2m. centres can be substituted, as indicated, in order to render boundary stockproof.

Dated the 31st day of October 1989.

Signed William Davies & Rees

Agricultural Valuers

Address Dolanog,
Welshpool,
Powys.

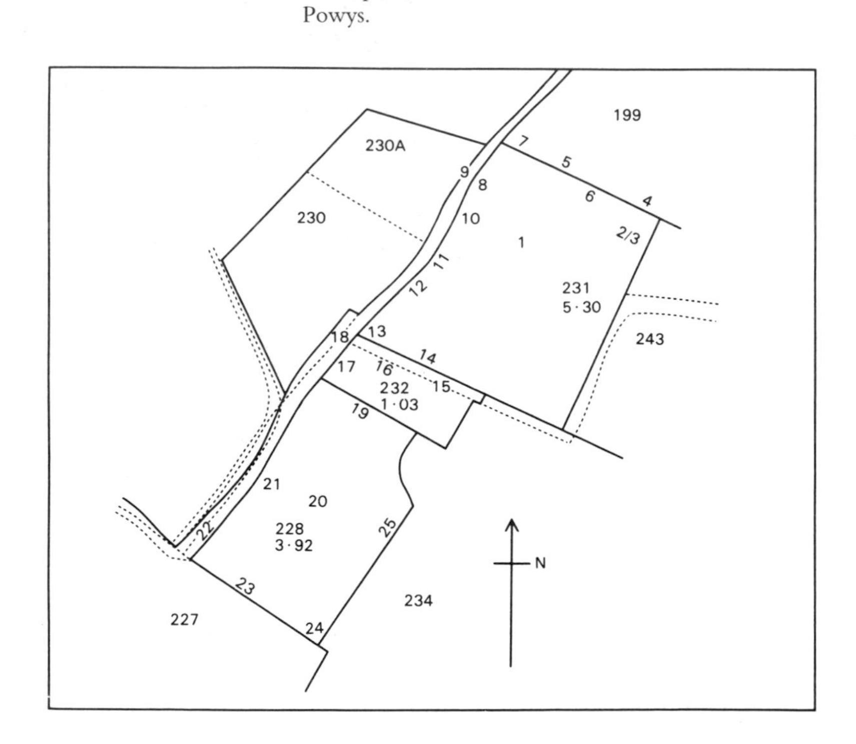

NOTES

In these Notes 'the Order' means the Agricultural Holdings (Arbitration on Notices) Order 1987 (S.I. 1987/710)

What to do if you wish:

(*a*) to contest your liability to do the work, or any part of the work, required by this Notice to remedy (Question (a)); or
(*b*) to request the deletion from this Notice to remedy of any item or part of an item of work on the ground that it is unnecessary or unjustified (Question (b)); or
(*c*) to request the substitution in the case of any item or part of an item of work of a different method of material for the method or material which this Notice to remedy would otherwise require to be followed or used (Question (c)).

1. Questions (*a*), (*b*) and (*c*) mentioned in the heading to these Notes can be referred to arbitration under article 3(1) of the Order. To do so you must serve a notice in writing upon your landlord within one month of the service upon you of this Notice to remedy. The notice you serve upon your landlord should specify:
   (*a*) if you are referring question (a), the items for which you deny liability,
   (*b*) if you are referring Question (b), the items you wish to be deleted,
   (*c*) if you are referring Question (c), the different methods or materials you wish to be substituted, and in each case should require the matter to be determined by arbitration under the Agricultural Holdings Act 1986 (c.5). You will not be able to refer Question (*a*), (*b*) or (*c*) to arbitration later, on receipt of a notice to quit. This action does not prevent you settling the matter in writing by agreement with your landlord.

Carrying out the work

2. If you refer any of these Questions (*a*), (*b*) and (*c*) to arbitration, you are not obliged to carry out the work which is the subject of the reference to arbitration unless and until the arbitrator decides that you are liable to do it; but you MUST carry out any work which you are not referring to arbitration.
3. If you are referring Question (*a*) to arbitration you may if you wish carry out any of the work which is the subject of that reference to arbitration without waiting for the arbitrator's award. If you do this and the arbitrator finds that you have carried out any such work which was not your liability, he will determine at the time he makes his aware the reasonable cost of any such work which you have done and you will be entitled to recover this from your landlord (see article 8 of the Order). This provision does not apply in the case of work referred to arbitration under Question (*b*) or Question (c).

What to do if you wish to contest any other question arising under this Notice to Remedy

4. If you wish to contest any other question arising under this Notice other than Question (*a*), (*b*) or (*c*), you should refer the question to arbitration in either of the following ways, according to whether or not you are also at the same time referring Question (*a*), (*b*) or (*c*) to arbitration:

(*a*) if you are referring Question (*a*), (*b*) or (*c*) to arbitration, then you must also refer to arbitration at the same time any other questions relevant to this Notice which you may wish to dispute. To do this, you should include in the Notice to your landlord referred to in Note 1 above a statement of the other questions which you require to be determined by arbitration under the Agricultural Holdings Act 1986 (see article 4 (1) of the Order).

(*b*) if you are not referring Question (*a*), (*b*) or (*c*) to arbitration, but wish to contest some other question arising under this Notice to remedy, you may refer that question to arbitration either now, on receipt of this Notice, or later, if you get a notice to quit. To refer the question to arbitration now, you should serve on your landlord WITHIN ONE MONTH after the service of this Notice to remedy a notice in writing setting out what it is you require to be determined by arbitration under the Agricultural Holdings Act 1986 (see article 4 (2)(a) of the Order). Alternatively, you have one month after the service of the notice to quit within which you can serve on your landlord a notice in writing requiring the question to be determined by arbitration under the 1986 Act (see article 9 of the Order). You will then have three months from the date of service of that notice in which to appoint an arbitrator by agreement or (in default of such an agreement) to make an application under paragraph 1 of Schedule 11 to that Act for the appointment of an arbitrator. If this is not done your notice requiring arbitration ceases to be effective (see article 10 of that Order).

Warning

5. Notes 1 to 4 outline the only opportunities you have to challenge this Notice to remedy.

Extensions of time allowed for complying with this Notice to remedy

6. If you refer to arbitration now any question arising from this Notice to remedy, the time allowed for complying with the Notice will be extended until the termination of the arbitration. If the arbitrator decides that you are liable to do any of the work specified in this Notice to remedy, he will extend the time in which the work is to be done by such period as he thinks fit (see article 6 (2) of the Order).

Warning as to the effect which any extension of the time allowed for complying with this Notice to remedy may have upon a subsequent notice to quit.

7. If your time for doing the work is extended as mentioned in Note 6 above, the arbitrator can specify a date for the termination of your tenancy should you fail to complete the work you are liable to do within the extended time. Then, if you did fail to complete that work within the extended time, your landlord could serve a notice to quit upon you expiring on the date which the arbitrator has specified and the notice would be valid even though that date might be less than 12 months after the next term date, and might not expire on a term date. Arbitrator cannot, however, specify a termination date which is less than 6 months after the expiry of the extended time to do the

work. Nor can he specify a date which is earlier than would have been possible if you had not required arbitration on this Notice to remedy and had failed to do the work (see article 7 of the Order).

Example Tenants Notice requiring Arbitration

AGRICULTURAL HOLDINGS ACT 1986 SCHEDULE 3, PART I, CASE D

AGRICULTURAL HOLDINGS (ARBITRATION ON NOTICES) ORDER 1987 (S.I. 1987 No. 710)

Re: The holding known as Part of Lower Barn Farm, Great Onen, Monmouth, Gwent—O.S. Nos. 231, 232 & 228

To: Mr. J. H. Fitzherbert, Landlord
c/o his agents—Williams, Davies & Rees
Dolanog
Welshpool, Powys

I HEREBY GIVE YOU NOTICE as follows:

I wish to contest my liability under the terms or conditions of my tenancy to do the work specified in the Notice to do work dated 31st October 1989 which you have given me in respect of the following items:

1)	O.S. 231	Pasture Foul. Eradicate couch, agrostis, docks, thistles, chickweed to bring field back into full production.
2)	O.S .231/243	Nettles, bracken and briars encroaching into field. Eradicate and control to bring area affected back into proper production.
3)	O.S. 231/243	Weak Gappy Hedge. Cut and lay and render stockproof.
4)	O.S. 231/199	Ditch overgrown and not cleaned out. Remove overgrowth, shrubs and briars and clean out to proper falls.
5)	O.S. 231/199	Weak, gappy, overgrown hedge. Cut and lay approx 17 yards of hedge, provide and erect a post and wire fence to the remainder to render the boundary stockproof.
8)	O.S. 231/ Roadway	Weak, gappy, overgrown, hedge. Cut and lay approx 50 yards of hedge, and provide post and wire stockproof fence to the remainder.

9)	O.S. 231/ Roadway	Ditch overdrawn and not cleaned out. Dig and clean out ditch to proper falls.
13)	O.S. 231/ Roadway	Barbed wire attached to Oak Tree. Remove.
16)	O.S. 232	This pasture field is currently in Arable Production contrary to its scheduling on the tenancy agreement. Return to pasture and in doing so eradicate couch, agrostis, buttercups, mayweed and also other arable weeds present.
20)	O.S. 228	Stubble Foul. Eradicate and control couch, agrostis, docks and buttercups and other weeds.
21)	O.S. 228/Lane	Encroachment of Hedge and Scrub Woodland into field. Cut and burn some 20 yard width of thorn bushes. Saplings (mainly ash), deadwood, brambles, bracken, thistles and nettles, eradicate and control to bring area back into proper production.

The grounds on which I deny liability to do such work in respect of the aforesaid are as follows:

Item No. 1 — I deny that this pasture is foul and infested with the weeds alleged any more than expected of an old pasture of this age and kind. The proper solution would be to plough up and re-seed but I am prohibited by Clause 10 (4) of my tenancy agreement from doing this.

Item No. 2 — I deny that the alleged infestation exists. There is no loss of production. It is practically impossible to eradicate the type of infestation alleged, if proved that it does in fact exist.

Item No. 3 — This hedge does not belong to the property I rent. This has always been maintained by the neighbour Mr. A. N. Other.

Item No. 4 — This ditch is not my responsibility and belongs away. In any event this ditch does not require cleaning out. The sides have been recently cut back by me and I always do this every year, in my own interest, but without accepting any liability to do so.

Item No. 5 — This boundary belongs away and has never been maintained by me during my tenancy.

Item No. 8 — This hedge is not ready for laying. It has been left to grow for laying in about 4 years time when it will be of sufficient growth. The boundary is stockproof and there is no need to provide the post and wire fence requested. Furthermore, I have absolute discretion as to which hedge I cut and lay each year. I am only required to cut and lay a 'proper proportion' and this has been done.

Item No. 9 — This ditch does not require cleaning out. It was cleaned out last year by a specialist contractor. The sides have recently been trimmed.

Item No. 13 — This barbed wire was not attached by me to the oak tree and existed as such when I took the tenancy.

Item No. 16 — Permission was granted by the previous owner to plough this field (copy letter dated 16th August 1960 attached) I am not required to re seed the same. It is denied that a serious infestation of weeds exist, particularly couch. The other weeds will be dealt with at the appropriate time in carrying out normal husbandry.

Item No. 20 — It is denied that the stubble is foul except that a minor infestation of annual weeds exists and these will be dealt with at the appropriate time.

Item No. 21 — I have no liability for reclaiming this area. It has always existed in this state and was so at the commencement of my tenancy. I have an old O.S. Map dated 1930 which shows it as rough.

I request all the above items to be deleted from the notice.

I also wish to contest the following questions on the Notice to do work dated 31st March 1989 which you have given to me in respect of the following matters:

(*a*) my liability, if any, to do the work stipulated and also the manner in which you prescribe it should be done.
(*b*) The enforceability of the Items 1–5 of the Rider forming part of your Notice.
(*c*) as to whether the period of 12 months given to carry out the work is a reasonable period. In my view it is not and is far too short a period in which to carry out all, or any, of the work stipulated.

I require the aforesaid questions and matters to be determined by Arbitration under the 1986 Act.

Dated this 10th day of November 1989.

Signed: A. Charles (Tenant)
....

Great Onen Farm,
Monmouth,
Gwent.

PRACTICE NOTE—Although the example reply does not show this, it is prudent when giving a notice requiring arbitration to oppose EVERY item. If this is done, items that are agreed, can be conceded at the hearing (if the Tenant has not in the meantime done them) and the Arbitrator can include the same in his Award or if the work is done, the Landlord can withdraw on those particular items. If Arbitration is not asked for on every item the work stipulated must be done.

CHAPTER FIFTEEN
Probate Valuations

15.1 GENERAL

Agricultural probate valuations usually cover both freehold agricultural properties, live and dead farming stock, and also quotas, where the deceased owned the same at the date of death.

15.2 FREEHOLD PROPERTIES

The valuation should be as comprehensive as possible and fully describe the property in detail, complete with a schedule of O.S. Nos. and areas, and have full particulars of outgoing charges. All separate parts should be adequately described including houses, cottages, buildings etc. If the farm is let, full details of the tenancy, term date, rental reserved (and when last revised) and repairing and insuring liabilities should be given. Also, if the tenants or occupiers (where they are not the same as the deceased) have carried out improvements (whether by consent or not) these should be carefully listed, as the value of these improvements are to be disregarded since they do not belong to the deceased's estate (if valuing for the Land-owner's estate).

If the property is in joint ownership, a 10% deduction should be made after dividing the share (if in the ownership of more than two persons, up to 15% can usually be deducted).

15.3 LIVE AND DEAD FARMING STOCK

Here, it is essential that everything should be described properly and individual values placed on each animal and item of produce, crops and dead farming stock. All animals should be described by their breed, age, etc., and valued accordingly. Dairy cows should be valued individually, with names or numbers given, as should be rearing cows with calving and service dates given. Store cattle and sheep can be valued in matching bunches. It is always important, to give, where possible, the ages of animals valued. A separate value should be given on each item of machinery, tractors, tools etc. In the case of important and valuable equipment, the year of manufacture or registration (if new, then) should be given.

This is not work that should be skimped and should be dealt with great care and in detail, as otherwise endless queries will arise from the Inheritance Tax Office.

15.4 RELIEFS

In forwarding the valuation to the deceased's Solicitor, it is prudent to remind him that, where the requirements are met, deductions of 50% of the value, as working farmer relief, should be made where the property is valued on a vacant possession basis, and 30% relief when the property is let. Also, 50% Business Asset relief from the value of the live or dead stock.

15.5 EXAMPLE OF PROBATE VALUATION

VALUATION FOR PROBATE

In the Estate of

FREDERICK ARNOLD Deceased

late of

BANK FARM, SCOTFORD, HEREFORD

Date of Death: 16th day of December, 1989
Date of Valuation: 10th day of January, 1990

A) FREEHOLD PROPERTY

BANK FARM, SCOTFORD, in the County of Hereford & Worcester

An agricultural investment comprising Farmhouse, a detached service cottage, farmbuildings and pasture and arable land, together with a small area of woodland, extending to 208.87 acres.

The holding is situated adjoining a council maintained road on the Scotford to Pearton road and is about 5 miles from the City of Hereford.

(a) The Homestead

The Farmhouse is constructed of brick, stone, part half timber and has a slate and clay tile roof. It is thought to date from the mid-eighteenth century. It contains:

On the Ground Floor:

Entrance Hall

Two Reception Rooms

Dining Room

Kitchen with Aga Cooker and Boiler

Scullery

Dairy

Washhouse

On the First Floor:

Landing

Five Bedrooms—3 double; 2 single

Bathroom: with heated linen cupboard

Separate W.C.

Boxroom

On the Second Floor:

Landing

Three Attic Rooms

This floor is disused

Outside: Large garden with small orchard

The Farmbuildings

These are convenient to the residence and comprise:

Steel framed concrete block and asbestos Loose Housing

Covered Cattle Yard 90′ × 60′ with raised centre feeding passage.

Dairy Complex: (now disused) comprising 8/16 Herringbone Milking Parlour with Milk Room adjacent.

Circular concrete Collecting Yard.

Steel, asbestos and concrete Silage Barn.

Traditional stone built and slate roofed Barn with 5 bay open sided cattle shed fronting an open yard.

Brick and asbestos Implement Shed/Garage.

(b) The Cottage

This is a detached brick and slated service cottage known as Stopforth Cottage and is situated off the entrance drive. It is on two floors and contains:

Hall

2 Living Rooms

Kitchen

Scullery

On the First Floor:

Landing

Three Bedrooms

Bathroom

Outside:

Garage and Garden.

(c) The Land

This is in a ring fence and is mainly level or gently undulating. The soil is a light, free draining loam, derived from the old red sandstone formation and is comprised in 10 enclosures, together with a small coppice of 4½ acres of young ash, oak and pulpwood. The land is watered with a number of automatic water troughs from the mains supply and also an intersecting stream. The whole extends to 208.87 acres as contained in the following Schedule taken from Herefordshire 1/2500 O.S. Sheet L. 10/11.

O.S. No.	Description	Area
782	Homestead	1.01
654	Pasture	24.65
821	Pasture	20.10
Pt863	Pasture	18.76
158	Arable	15.19
242	Arable	12.92
111	Pasture	27.98
515	Pasture	24.81
992	Coppice	4.56
002	Pasture	18.76
004	Roadway	1.04
182	Pasture	21.11
259	Pasture	17.98
		208.87 Acres

(d) Tenancy

The holding is let on an annual candlemas agricultural tenancy to Mr. J. Fraser-Wood by virtue of an agreement dated 10th February 1975 the tenancy having commenced on 2nd February 1975. Repairing liabilities are in accordance with S.I. 1973 1473. The rental currently payable, with effect from 2nd February 1988, is £8,000 per annum.

The tenant has a son, working full time with him, aged 28 years.

(e) Tenants Improvements

The tenant, at his own expense, has carried out the following improvements.

i) Modernised the farmhouse include provision of hot water system, bathroom (complete with all fittings) installed Aga Cooker and provided all kitchen equipment.
ii) Constructed the farm drive including finishing with an asphalt surface.
iii) Tile drained O.S. Nos. 821 & 182 (30 Acres)
iv) Installed the whole of the dairy complex comprising Herringbone Parlour, Milk Room, Collecting Yard, Covered Cattle Yard (90′ × 60′) and the Silage Barn.

Landlords written consent for all this work has been given, depreciating the cost of each item down to £1 over a period of 15 years.

(f) Outgoings

Drainage Rate	£58.70
Charity Rent	£10.25

(g) Quotas

The holding has a wholesale milk quota of 292,061 litres, of which, it is considered, informally, the deceased owned 70%.

(h) Services

Mains electricity and water connected.
Drainage of residence and cottage to individual septic tanks.

(i) Valuation

Subject to the existing tenancy and including 70% of the milk quota, but dis-regarding the value of the tenants improvements £200,000 (Two Hundred and Eighty Thousand Pounds).

B) LIVE & DEAD FARMING STOCK at MANOR FARM, SCOTFORD which the deceased occupied as an agricultural tenant.

(a) Cattle

41 20 month old Steers viz:		
25 Hereford cross Fresians @ £480	£12,000	
15 Charolais cross @ £510	7,650	
1 Shorthorn cross	420	
20 Hereford cross Friesian 15 month steers @ £360	7,200	
61 Total		27,270

(b) Sheep

184 Breeding Ewes viz:	
105 Suffolk cross 3—4 year old @ £38	3,990

61 Mule 2 year old @ £60	3,660	
18 Mule Stock Ewes (BM) @ £32	576	
2 Suffolk 2 year old Rams @ £80	160	
186 Total		8,386

(c) Pigs

10 Large White Cross Landrace Sows with 91 piglets (up to 6 weeks old) @ £240	£2,400	
4 Large White Cross Landrace Sows due to farrow within 1 month @ £150	600	
10 Large White Cross Landrace Sows served 1 month to 2½ months @ £125	1,250	
5 Large White Cross Landrace Sows recently weaned @ £100	500	
1 Large White Stock Boar	200	
80 Pork weight pigs @ £55	4,400	
70 Medium Store pigs @ £35	2,450	
35 Weaners @ £22.00	770	
306 Total		12,570

(d) Agricultural Equipment

1974 Commer Stocklorry with 24′ Box Reg. No. AHN 007M121,600 recorded miles	£ 600
1982 M.F. 595 Tractor Reg. No. HGV 116	3,100
1976 M.F. 135 Tractor Reg. No. SHJ 765	1,700
1978 M.F. 187 S.P. Combine Harvester 10′ cut Reg. No. RTM 112R	3,000
Urry Cattle crush	140
Circular Cattle Feeder	40
6–8′ metal cattle racks @ £35	210
Wooden Cattle Rack	25
4 Covered Sheep Racks @ £50	200
1 Covered Sheep Rack (Very poor order)	10
10 Sheep Troughs viz:	
8 @ £10	80
2 (very poor) @ £2	4
Fraser 8T. Grain Trailer	1,500
F.R. 5 Furrow Plough	500
Fahr K.M. 22 Mower	250
Hay Bob	150
M.F. 3T. Trailer	200
I.H. Fore-end Loader with Fork, bucket and bale spike	525
Parmiter Silage Grab	400
Scrap Iron	10
Small Tools	30
B & D Electric Drill	25
M.F. 30 G & F Drill (3 m)	850

I.H. 440 Baler	400	
Howard 150 Rota Spreader	450	
Vicon 3 m Power Harrow	1,200	
Cambridge Gang Rollers (18′)	425	
Parmiter 13′ zig-zag Harrows	200	
		16,224.00

(e) Produce (Consuming Values)

Hay

5,500 bales 1989 baled seeds hay (very weathered) = 110 tonnes @ £40	4,400	
225 tonnes Silage @ £12	2,700	
25 tonnes 1989 baled Wheat Straw @ £15	375	
		7,475.00

(f) Growing Crops

O.S. Nos. 104 & 2002–20 Acres Avalon Winter Wheat @ £74.23 per acre (cost of seeds, fertilisers, cultivations & enhancement value)		1,484.60

(g) Tenants Pastures

O.S. No. 12 17.11 acres (4 yr. ley sown 1986) 17 acres @ £15	225	
Pt. O.S. No. 87 20 acres (L.T. Ley sown 1987) @ £40	800	1,055.00

NOTE: No valuation has been made of any remaining improvements (short and long term) or unexhausted or residual manurial values since it is considered that these will offset a dilapidations claim that will arise on the termination of the tenancy.

SUMMARY

	£
(a) Cattle	27,270.00
(b) Sheep	8,386.00
(c) Pigs	12,570.00
(d) Agricultural Equipment	16,224.00
(e) Produce	7,475.00
(f) Growing Crops	1,484.60
(g) Tenants Pasture	1,055.00
Total	£74,464.60

GRAND SUMMARY

	£
FREEHOLD PROPERTY	200,000.00
LIVE & DEAD FARMING STOCK	74,464.60
	£274,464.60

We, the undersigned, having inspected the above, DO VALUE, the same, at the date of death, in the figures given.

.... F.R.I.C.S. F.S.V.A
F.A.A.V.
of Messrs. Williams, Davies & Rees
Agricultural Valuers
Dolanog
Welshpool
Powys.

CHAPTER SIXTEEN

Stock Taking or Income Tax Valuations

16.1 These valuations generally form an important part of any agricultural valuation practice. The work, system wise, is repetitive and most valuers use special valuation proformas for completion with numbers and the detailed calculations. Adopting this system means that items are not forgotten and the valuation is done in a standard and consistent way each year. (See Example Attached).

16.2 Livestock should be valued at cost of production or market value, whichever is the lowest. In the case of homebred stock, the cost of production can be taken as follows:

Cattle 60% of Market Value
Other Livestock 75% of Market Value

16.3 In the case of livestock *purchased and reared on*, the valuation should be the *higher* of the cost and 60% of market value of cattle and 75% of the value of all other livestock. Check that the cost does not exceed the market value and value on the lower figure.

Harvested crops such as corn, potatoes, sugar beet etc., should be valued at lower of cost and net realisable value. Calculate cost of production to determine cost. Many farmers do not keep records and where this is so, it has been agreed between the Revenue and the N.F.U. that 85% of the market value on the holding on the date of valuation can be adopted.

Hay, straw and silage, produced on the farm, should be valued at cost of production or at 60% of market value where no cost figures are available. Purchased fodder should be valued at lower of cost and net value.

In the case of growing crops these are valued at cost of seeds, cultivations, sprays and fertilisers. Enhancement value is not included. Any fertilisers applied to pastures and other acts of husbandry such as orchard spraying should be included at cost.

16.4 Note, where the production herds (this only applies to milking and rearing cows, bulls, breeding ewes, rams, sows and boars) are on the HERD BASIS no value should be given for these but the actual numbers should always be shown on the valuation. Note, the Herd Basis is only applicable to *production stock* and not to first calving heifers, in-pig gilts, or in-lamb hoggetts which at the date of valuation have not done any 'producing'. If the ewe flock has lambed, and the flock is on the 'herd' basis, only the lambs are valued and not the ewes. Similarly, where there is a suckler herd, again on the 'herd' basis, only the calves are valued.

16.5 It should be borne in mind that this is a 'stocktaking' valuation comprising the livestock, fodder, produce, stores, growing crops, acts of husbandry and fertilisers applied on a farm. It does not embrace equipment, fencing, and drainage works. Very few valuers include unexhausted manurial values, unexhausted lime and residual value of feeding stuffs. These are not usually charged for when an owner occupier vacates a farm and where a tenant vacates—he might well have a dilapidations claim against him and these items, in a tenants case, would serve as a partial offset against such a claim.

16.6 EXAMPLE OF VALUATION PRO-FORMA

INCOME TAX VALUATION

NAME .... VALUATION AS AT ....

ADDRESS .... DATE OF VALUATION ....

....

	Details	@	£	p	£	p
CATTLE						
Milking Cows						
Rearing Cows						
Bull(s)						
....						
In-calf Heifers						
Two Year Old						
18/20 months						
12/15 months						
6/10 months						
2/5 months						
Young Calves						
....						
....						
SHEEP						
Ewes in/and with lambs viz:						
....						
....						
....						
Rams						
....						
Lambs						
Cull Ewes						
Wethers						
HORSES						
....						
PIGS						
In-farrow Sows						
Breeding Sows						
Breeding Sows and						
Litter(s)						
Boar(s)						
In-farrow Gilt(s)						
Breeding Gilt(s)						
Weaners						
Porkers						
Baconers						
....						
POULTRY						
Hens						
Chicks						
Ducks						
Turkeys						

	@	£	p	£	p
				Forward	
FEEDING STUFFS					
T					
T					
T					
HAY					
T					
STRAW					
T Barley Straw					
T Wheat Straw					
T Oat Straw					
SILAGE					
T					
ROOTS					
T Mangolds					
SEEDS & SEED CORN					
....					
....					
SEED POTATOES					
T					
T					
WARE POTATOES					
T					
FERTILISERS IN STOCK					
T					
T					
T					
AGRICULTURAL STORES					
....					
....					
....					
FUEL					
L.					
L.					
GROWING CROPS & ACTS OF HUSBANDRY					
Acres Wheat					
Acres Oats					
Acres Barley					
Acres					
....					
Acres					
....					
Acres Potatoes					
....					
Acres Swedes					
Acres Kale					
Acres Sugar Beet					
Acres Young Seeds Planted					
....					
....					
Manures & Lime Applied:					
....					
....					
....					

16.7 EXAMPLE INCOME TAX VALUATION

FAIR OAK FARM

HENDLEFORD

INVENTORY
AND VALUATION

of the Live and Dead Farming Stock, Produce, Fertilisers, Stores, Fuel, Growing Crops and Acts of Husbandry
the property of M. J. L. Yeoman
as at 25th March 1990

CATTLE:	
78 Milking Cows 6 Rearing Cows 1 Bull } HERD BASIS	
74 Other Cattle	£20,560.00
SHEEP	
172 Ewes 4 Rams } FLOCK BASIS	
299 Other Sheep	3,837.00
FEEDING STUFFS	3,864.00
HAY	2,130.00
STRAW	672.00
SILAGE	4,180.00
FERTILISERS IN STOCK	1,955.00
AGRICULTURAL STORES	197.60
FUEL	208.00
GROWING CROPS & ACTS OF HUSBANDRY	9,717.20
	£47,320.80

.... A.R.I.C.S. A.S.V.A. F.A.A.V.
of Williams, Davies & Rees
Chartered Surveyors and Valuers, Dolanog, Welshpool, Powys

CHAPTER SEVENTEEN

Factors to Consider in Valuation of Freehold Farms and their Rental Values

17.1 GENERAL

Valuation of agricultural freeholds and their rental value is a skill only acquired after a number of years' experience, and the young valuer will have an advantage if he has actually worked on a farm and will learn of the difficulties or advantages of any particular farm. There are many factors to consider and these include:

17.2 QUALITY OF THE SOIL

The productivity and thus profitability of any farm (except an intensive production unit such as a pig farm) depends on the quality of the soil. The ideal is a fairly easily worked soil, preferably deep, free draining level loam (but not too sandy as it dries out). It should be versatile to the extent almost any crop can be grown on it, and should be ideal for all arable crops as well as livestock enterprises dependent on grass, such as dairying.

Frequently, much level land is clay based and thus heavy and often water-logged. Potatoes and sugar beet cannot be satisfactorily grown on heavy land because they cannot always be harvested in the wet conditions experienced in the autumn. Similarly, it may sometimes prove difficult to plant such land with autumn sown cereals. This type of soil usually has a high water table, but in dry conditions tends to dry out very quickly and large cracks develop. Examples of this soil are to be found extensively in Worcestershire and also the Severn Vale in Gloucestershire. This soil is also cold and late in the spring.

An ideal soil is often an alluvial loam of good depth with little stone content. This is generally relatively free-draining, does not get water-logged and can grow most crops. However, this type of soil is normally found in valley bottoms, some of which are prone to flooding.

Sandy soils, whilst easy to work, dry out very quickly and unless the growing season and the summers are wet, yields suffer by comparison with crops grown on rich loamy soils. Sandy soils also wear out ploughs and cultivating equipment and tyres because they are so abrasive. However, these soils have the advantage of being early in the spring and are also good for late summer and autumn growth. Generally, it should be possible to harvest sugar beet and potatoes, and also to plant autumn sown crops when there may be difficulty on heavier soils. Sandy soils benefit considerably where irrigation facilities exist. They also tend to 'blow' in certain situations, particularly in the Eastern Counties where there are few windbreaks and established vegetation. The yields are seldom as good on light soils as on heavier soils. In particular, pastures do not grow as thick and grass plant population is usually less. Sandy soils are often 'hungry soils' and frequently are deficient in lime, phosphates and potash which tend to leach easily, particularly on banky fields. The soil on banky fields also tends to wash out easily during heavy rainstorms.

Many soils, particularly at higher elevations in limestone areas, are very stony and rocky. Often there is little depth of soil and the underlying rock formation is near the surface. Crops suffer in drought periods and sometimes completely die off or do not develop. These shallow soils are usually only suitable for grazing but, nevertheless in the last two decades have been ploughed. Cereals are often the only arable crops that can be satisfactorily grown, since they have so little depth. (often only 4″ or less top soil). High value cash crops, such as potatoes and sugar beet, cannot be satisfactorily grown as these need deep soils, and the presence of stone and rock does not allow mechanical harvesters to function properly. Stony and rocky soils 'burn' in periods of drought and are also very damaging to tractor tyres, and implement breakages and damage are frequent. Examples of such soils are to be found in the Forest of Dean, the Cotswolds, Pennines and the flinty areas of the Southern Counties. The existence of stone boundary walls is the first indicator of the presence of stone, apart from the rock outcrop which sometimes are visible. Trees are usually stunted and hedges are sparse and often are only a few clumps here and there.

Sometimes clay patches, which are heavy and wet, and also areas with rising springs, causing wet patches, are found in all classes of land, even in sandstone soils. These as a rule need draining.

The presence of thick continuous hedgerows, well grown, and also good large trees is usually an indicator that the soil is deep and of good quality. The presence of elm trees (although now devastated by Dutch Elm disease but elm hedges still exist) is an indication of good, deep, rich soils.

Ditches and watercourses are seldom found on free-draining soils, particularly in sandy soils, chalk soils and limestone areas. The exception is in the valleys and river flood plains where the soil tends to be heavier, and is often clayey loam. The absence of ditches can be an advantage, since they do not have to be cleaned out, but may be also a disadvantage if there is no other water supply available for stock.

Some soils overlie impervious clay strata and unless drained are perpetually water-logged, unproductive and grow, in particular, rushes and sometimes reeds. Drainage is not always effective, especially if the area is very low lying. Contrast, the Fenlands which were made into some of the most productive arable land in this Country by drainage despite, in many areas, being below sea level. Very deep drainage channels and levels (Bedford Levels) were dug and the water pumped out into the sea. The Fens are of a peat soil which is inherently rich and frequently capable of producing two crops a year. Here, in this area of comparatively low rainfall, the presence of a fairly high water table is an asset.

17.3 THE HOMESTEAD

The existence of a good quality residence has an important bearing on farm values, now that the values of residences has escalated. Very often, when farms are sold, the residence gardens and a small paddock is lotted separately and is often purchased by a non-farming buyer.

It is very advantageous to have good modern buildings, particularly well designed, soundly built and properly maintained. Specialist buildings such as poultry and pig units do not appeal to every type of buyer and, if extensive, can be a hindrance to a sale, particularly if there is a slump, say, in egg or broiler profitability or the pig industry, as occasionally occurs.

Dairy buildings, especially if well designed with a milking parlour, cubicle housing and silage pits, on smaller farms (say up to 200 acres) are an asset to a sale, since many farmers who farm holdings of this size are dairy farmers. If they did not exist it is a costly matter to provide a modern dairy set up. A milk quota, is of course, essential, to practice dairying but purchasers of holdings without quota can purchase quota if they intend to practise dairying and where no, or insufficient quota exists.

The most popular and useful buildings are portal-framed general purpose buildings which can be used for housing livestock, whether beef or dairy cattle, used for sheep housing, storage of grain, potatoes, hay and straw or farm equipment. Such buildings should have high headroom (minimum 16′ at eaves), (low headroom buildings are unhealthy for cattle as they frequently get virus pneumonia attacks), and be maintenance free—e.g. precast concrete and galvanised steel which does not rust and does not require painting. Also, if the building is a covered silage pit it must have a high headroom, as tractors with safety cabs cannot otherwise have enough headroom to operate when consolidating the silage.

The existence of good modern grain stores and potato stores (both preferably with underfloor ducting—in the case of grain stores for ease of loading with grain buckets and in both cases, for drying) on an arable farm is obviously an asset.

The presence of adequate farm cottages on larger holdings, where they are needed, is a further asset.

If the farmbuildings on any given holding are old, traditional buildings, (and with no modern buildings) this will, in some cases, result in the capital value being a good deal less than when a farm is suitably equipped with modern buildings. However old stone buildings which can be converted into dwellings or holiday lets, provided they are a suitable distance from the residence, are nowadays an important plus factor, in the value of any holding.

Many farms equipped with modern buildings, have been badly designed, of a poor layout, and sometimes such buildings are too specialist in nature, and in some cases these buildings are now outdated. Examples are: buildings which have too low a headroom, cattle buildings with badly designed feeding and bedding arrangements (cattle cannot be fed by tractor and forage wagon, and cattle cannot easily get access to silage clamps); slatted floor buildings (not every farmer wants to keep his cattle on slats and the buildings really cannot be used for anything else); cattle buildings with tower silos and automatic feeding of tower silage systems (not everyone wants to be dependent on this system as if there is a power cut the cattle cannot be fed, apart from the numerous mechanical breakdowns that occur); concrete floors not damp-proofed and where grain cannot be stored owing to rising dampness.

It follows, therefore, that where a holding is inadequately equipped, providing it can be purchased at a lower figure to reflect the cost of providing essential buildings, this can be beneficial since exactly what is needed can be erected.

17.4 SITUATION

The location of a farm has a major bearing on values and the following matters are significant.

17.4.1 Position in relation to towns

A holding convenient for towns, markets, with shops, schools and other facilities, is usually worth more than a holding that is remote. Transport costs time and money. Conversely, if a farm is on the outskirts of a town, this can be a nuisance because of trespass, vandalism, and often a sheep flock cannot be kept owing to dogs worrying sheep. Also, a number of footpaths usually exist on holdings near towns and these are often much used and result in further trespass.

17.4.2 Position in relation to roads

A farm with the homestead immediately adjacent to and with *direct access* from a much used road can be a dangerous place to live. Fast moving traffic is a major hazard to the movement of tractors and equipment and also livestock. Where some land exists across a road and a dairy herd is kept, the movement of cows four times a day for six months of the year across this road is a positive hazard. Vehicles run into stock and kill or injure them, and every time stock is taken across two or three persons are needed to warn traffic and drive the stock.

Coversely, a farm situated with its homestead some distance, say up to 500/600 metres of a main road and with no or very little land severed, has many advantages.

It is not always an advantage to have a farm approached by a long *narrow* council road as there are frequently no proper passing places, and vehicles have to proceed with caution. However, a little used road can sometimes be an asset, since it might give access to roadside fields and is, of course, particularly useful in wet weather to give firm access to these fields. Farms with their own long private drives leading off roads have the disadvantage that the drive is costly to maintain.

In this age of reduced farm profits a situation where a farm shop can be operated, or a Caravan Club Licence obtained can be a major asset as these can be profitable diversifications.

17.4.3 Farm approaches

A farm with a good level access has a distinct advantage over a holding approached up a steep hill, which can be difficult in winter. Likewise, if the access road is narrow and exposed it can easily become blocked with snow drifts. Occasionally, farms have low lying accesses which are liable to flooding, and this can be a problem at times.

17.4.4 Farms on flood plains

Valley land is usually good productive land, but if it is liable to flooding this can be a drawback since, in some valleys, the river floods on a number of occasions annually, mainly in the winter. This restricts its user and losses can arise to cereal crops, and it is frequently unsafe to grow potatoes and root crops, in particular sugar beet, since they cannot always be harvested. The odd summer deluge resulting in summer flooding can result in serious losses to growing crops. If a farm homestead is itself on a flood plain, and is susceptible to flooding, this can be a serious problem. Land on flood plains usually has a large number of ditches and water courses have to be cleaned out, and this can prove an expensive, recurring liability.

17.4.5 Elevation and aspect

A holding at a low elevation is generally more productive and often has a great deal more inherent fertility than a farm at a high elevation, of say 500 ft and upwards. The growing season is longer and the farm is generally much earlier than a high-lying farm, which is often (unless in a sheltered high valley) cold and late. However, low-lying farms in the Eastern Counties, being on wide open plains, tend to be cold and exposed. The growing season is also longer in the south of the country than further north, but the earliness of any holding usually depends on whether it has good free draining sheltered soil, and also its location. The influence of the Gulf Stream is an important factor and in western areas e.g. Cornwall, Pembroke etc. and where early potatoes are grown since some of these areas are frost free. Much low lying land is, however, wet and the soil is of clay and impervious and has a high water table, and is consequently late in spring e.g. the Somerset Wetlands. It is sometimes difficult to drain this wet land, particularly if very low lying, and drains, when laid, frequently get blocked up. The result is rush-infested, unproductive land, which can only be grazed for a limited period in the summer months.

The aspect of land is very important. If it faces north east, it is cold and late. Snow and frost clears last of all from land with these aspects, by contrast to land with a south and west facing aspect having plenty of sun. However, in very hot summers, north and east facing land does not 'burn' to the same extent, particularly where rock is near the surface. Land adjacent to woodlands, especially if it is on a north and east slope, is usually very unproductive, and it is rare for anything much to grow close to woodlands and under trees. Also, hedges do not grow well by woodland, and it is to be noted that rarely is a good hedge to be found under a tree or trees. Overhanging boughs can also be a nuisance in the use of machinery, particularly high tractors and combine harvesters.

Banky land is difficult land to farm. If the banks are severe, tractors with twin wheels (and four-wheel drive) or caterpillars are needed for reclamation, re-seeding, fertilising and thistle-cutting. However, all banks are dangerous, and considerable caution must be exercised with the use of any tractor and vehicle. Steep land is not generally suited for arable production and combine harvesters do not function well on banks as some of the grain harvested is discharged through the back of the machine. This type of land is generally most suited for grazing purposes. Banky fields have a greater surface area than that shown on the O.S. maps!! This is why many marginal land and hill farms carry greater headage of stock than one would expect them to keep.

17.5 FENCING, GATES, DITCHES AND DRAINS

A most important factor on any stock farm is the presence of proper stockproof hedges or fences.

17.5.1 Hedges

Good stockproof hedges are essential on stock farms. They serve to keep stock in the fields and also give shade from the sun and protection from wind and rain. This is why so many high hedges, which one may consider should be laid, exist on stock farms. It is also wise to allow a few hedgerow trees to grow on instead of trimming them as they provide shade and shelter.

Hedges grow well on rich loamy and heavy soils but do not thrive on thin soils. A hedge usually takes 10/12 years from the date of planting to the first laying. If a new hedge is planted, it should be of quickthorn, planted in two rows, and staggered.

17.5.2 Fences

The most suitable economic fence on a farm is medium/heavy gauge pig netting, erected on tanalised posts (erected every 2 m with adequate straining posts and struts at the end or at any change of direction). Two strands of barbed wire should be provided to provide a minimum height of 4 ft. Such fence should, if properly erected, last for 15/20 years, unless situated in an area of industrial pollution or near the seaside.

When holdings are converted to total arable farms, hedges and fences are often neglected and the only attention hedges have, as a rule, is trimming. If the farm, is at a subsequent time used as a livestock holding, it is a very costly and time consuming matter to erect permanent stockproof fences.

17.5.3 Gates

Are important on any mixed or stock farm. A well run farm should have good quality iron gates of adequate width (preferably 15′ wide) properly hung and with proper fasteners (not fastened with baler twine!). It is important that a gate, when erected, should have adequate strength and not be too light in construction. Cattle rubbing against a light gate will soon buckle it and will bend the top rail if they try to jump over when it has been hung too low.

17.5.4 Ditches and drains

Where these exist, should always be maintained and kept clean and running. They have usually been cut and laid for the purpose of draining the land, and if they are not maintained, do not serve the purpose. Drain outfalls should be kept open at all times. Ditches should be protected from livestock treading them in, with barbed wire fencing.

17.6 FARMING STANDARDS

Farming standards have much effect on farm values. Clearly, if a farm is farmed to a high standard and is highly productive, neat and tidy, it should be worth more than a comparable holding which has been neglected and needs considerable expenditure and effort to bring back into full and proper production. Often, the fixed equipment is neglected and it has no modern buildings. Again, the land often has been allowed to become weed-infested, impoverished and lacking in required levels of potash, phosphate and lime to sustain proper plant growth. Hedges, fences, ditches, gates, culverts, drains and roads may be neglected. The cost of improving all or any of these to proper standards in substantial and often it my take 3–5 years, or even more, to complete the work.

17.7 SERVICES

The availability of all main services is a great asset. However, some of these main services can be costly in use. If a dairy farm is entirely dependent on mains water supplies and a large herd is kept, water charges will be considerable. Private supplies, which cost very little or nothing avoid considerable expenditure. Likewise, the availability of irrigation facilities from streams and lakes has considerable value and can increase productivity dramatically despite the charges water authorities make for such. Effluent charges, where effluent is discharged into main sewers on dairy, pig and specialist farms on the outskirts of towns, can also be very high. If effluent can be treated on the farm, as is often the case, this will again avoid considerable expense.

17.8 SPORTING

The existence of good sporting on a farm, in particular well placed woods, coppices and dingles that provide cover for game and can create a good shoot, is of much value. If some of these woods are so placed that they can provide 'high' sporting birds, this will again improve the shoot. Likewise, the presence of duck flight ponds or small lakes that can be so developed is a great attraction. Even if a farm has all these facilities and no game, a good shoot can be created by rearing game and providing further cover by growing root crops, such as kale, swedes, and possibly leaving small areas of unharvested grain, and also planting small corners up.

If a holding has a trout stream or a lake which can be stocked or, better still, some river fishing (with good pools for holding the fish) this, again, is an attraction and an asset.

Sometimes, farms are sold with the shooting and other sporting rights reserved. This is a disadvantage, apart from the interference caused by persons exercising these rights.

17.9 DEVELOPMENT VALUE AND COTTAGES

There is sometimes a possibility that a farm may have some development potential if it has land adjacent to a town or village. Provided this is a distinct possibility and not a far distant hope, it will add value to a farm. Likewise, the presence of a number of cottages (provided they are not restricted in use to agricultural occupation) can be an asset, particularly as often these can be sold off at good prices even if derelict.

In recent years considerable value has been added to many farms which have barns and other redundant buildings that can be converted to dwellings and holiday lets.

17.10 WOODLAND

An area of woodland can be most useful on a farm quite apart from providing cover for game, shade and shelter and for use as a windbreak. Farmers are generally in need of timber for fencing posts, gate posts, etc. Material of this kind is expensive to buy and a good and suitable free farm supply is thus an asset, provided the area concerned is not disproportionate to the size of the holding. Also, it may be possible, in some cases, where the area concerned is not too steep, to clear and reclaim this for agricultural use. Since the advent, in the last decade, of woodburning stoves, an adequate free supply of firewood is of great value.

Generally, the most valuable timber for a farmer is hardwood such as oak and chestnut (for fencing posts), and the best softwoods (for rails and conversion), larch and spruce.

It is usually prudent for a farmer to plant unproductive areas, such as dingles and odd corners with quick-growing softwoods and wet areas with poplar. Apart from the timber value the shade and shelter provided, this will also be a conservation measure.

17.11 EASEMENTS, WAYLEAVES, RIGHTS OF WAY ETC.

Rights of way, easements including pipeline easements, electricity board lines, restrictions etc. can have a depreciating effect on capital and rental values and their effect should always be considered when valuing a freehold or assessing its rental value.

17.12 GENERAL

A few matters considered as a 'plus' to be of value on a farm are:

17.12.1 A good hard internal spine road, or road to give easy access to fields, and if in the form of a lane, is ideal for driving stock along without damaging crops.

17.12.2 Natural water supplies, such as an intersecting or bordering streams.

17.12.3 Good shade and shelter, which is essential on livestock farms.

17.12.4 That the farm has a good layout with a centrally placed homestead with not too much of a 'pull up' towards the homestead. Steep roadways of this kind can be difficult at almost any time, but particularly in the winter and, of course, are energy consuming.

17.12.5 That the homestead should, if at all possible be sheltered and protected from the prevailing wind. If it is very exposed, it will be a cold place at which to live.

17.12.6 The land not being too exposed to prevailing winds, e.g. a farm on a high altitude with no protecting hills or windbreaks.

17.12.7 The possibility of splitting a farm so that a portion can be sold off, if desired may well enable a potential purchaser to conclude a deal.

17.12.8 The quality of the soil always, in the final analysis, determines the value of a farm. Versatile soils are favoured, being not too heavy and not too light, but probably a deep loam with some 'body' in it i.e. tending to be slightly more heavy than light. Good land can be found everywhere, but some of the most fertile and popular agriculturally favoured areas of the country are Herefordshire, Shropshire, Taunton Vale, Cheshire, The Fens, Vale of York, Fife, Angus and Ormskirk area of Lancashire. The high prices realised for farms in these districts indicates their popularity.

The Agricultural Land Classification map should always be studied. To date, these maps are 'provisional' editions only and may yet be amended. The land is placed in 5 categories with Grade I and II the best, and with only minor limitations. Grade III is also good with only moderate limitations. Grade IV has severe limitations and Grade V is usually rough grazing. Generally speaking, these maps are reasonably accurate but they are on a small scale and deal with fairly large blocks. Often, some really good small areas of land are not indicated as they should be, and the converse also occurs.

17.12.9 It is now obvious that holdings possessing quotas, whether they are potato, sugar beet or milk quotas, will in the future have considerable attraction to farmer who is interested in producing a commodity for which the holding has a quota. This is best illustrated by stating that prior to the introduction of milk quotas, there was no restriction on milk being produced and sold off any farm provided all regulations were complied with. This will not be possible in the future, unless that particular holding has a milk quota. Quota (except beet) can be purchased but is expensive. The amount of the quota is also very relevant—clearly a holding with a large potato or milk quota will be more attractive and worth a great deal more to any buyer than similar holdings with small or no quotas. This has had far reaching effects on capital values and rental values.

17.12.10 The possibility of diversification, e.g. running a farm shop, caravan site, bed and breakfast business etc. etc.

CHAPTER EIGHTEEN
Inventories (Records)

18.1 The accurate preparation of inventories is important on the completion of any agricultural valuation. A good inventory is of particular importance to a tenant farmer when he enters a farm, and this can be very useful when he vacates. The inventory will prove what he paid for e.g. tenant's fixtures, tenant's pastures, improvements and also the cropping of some land on entry. The C.A.A.V recommend that inventories should always be prepared by the incomer's valuer as he is more likely to be concerned about its accuracy than the outgoer's valuer.

18.2 Whilst giving too much detail may possibly result in questions being asked and sometimes controversy between a client and his valuer, (this should not arise provided there is proper consultation between them), it is nevertheless important to set matters out in a fairly detailed manner, particularly:

(i) values given in respect of any equipment taken over. If this is not done, this information will have to be divulged at a later date for accountancy purposes. It is not necessary to give individual prices, except perhaps for tractors and major items of equipment.

(ii) details of tenant's pastures i.e. O.S. Nos. areas when laid down, and duration of the ley.

(iii) tenant's fixtures and improvements should be clearly identified.

(iv) where dilapidations are valued and offset, it is advisable to give as an annex to the inventory, details of what items have been allowed for and with the total compensation allowed, but not individual prices. This will show to the incomer what he has been allowed for and what he is expected to remedy.

18.3 It is not particularly important to state what the previous crops grown prior to those actually valued. The incomer can ascertain this from the outgoer himself, and rotations are more often than not no longer practised.

18.4 The inventory should give the effective date of the valuation, the date when the actual valuation was carried out, and should be signed by both valuers. The incoming tenant should keep the inventory in a safe place, preferably with the tenancy agreement, as it could well be needed one day.

18.5 SPECIMEN INVENTORY

INVENTORY & VALUATION

of

IMPROVEMENTS, TENANT RIGHT

FIXTURES AND EQUIPMENT

at

LOWER MOOR FARM, PENTLAND

in the County of GWENT

From: A. H. MOOR ESQ. Outgoing Tenant

To: J. H. LEWIS Ingoing Tenant

As at 2nd February 1990
Valuations taken and made 31 January 1990

1. PRODUCE

1.1 Hay —at consuming value
Five bays baled 1989 Seeds Hay in Dutch Barn
Part bay baled 1988 Meadow Hay in Wergins Barn

1.2 Straw —at consuming value
Two bays baled 1989 Winter Barley Straw in Dutch Barn
28 big bales 1989 Winter Wheat Straw in O.S. No. 123.

1.3 Silage —valued in accordance with the basis given in N.P. 154, C.A.A.V. (as amended)
Part used clamp, 1989 grass silage
Clamp 1989 Maize silage

1.4 Roots —Clamp Golden Tankard Mangolds

2. GROWING CROPS

(All at cost of cultivations, seeds and, where applicable, sprays and enhancement value)

2.1 Stubble Turnips	—O.S. No. 345. 3.56 ha The growing crop of stubble turnips on 2 ha
2.2 Swedes	—O.S. No. 365. 4.81 ha The growing crop of swedes on 2 ha
2.3 Winter Wheat	—O.S. No. 567. 11.71 ha The growing crop of Avalon Winter Wheat
2.4 Winter Barley	—O.S. No. 789. 13.45 ha The growing crop of Igri Winter Barley

3. CULTIVATIONS

3.1 Ploughing	—Ploughing in O.S. No. 891. 10.2 ha Ploughing in O.S. No. 894. 11.21 ha

4. LABOUR TO FARM YARD MANURE

Heap carted to O.S. No. 788

5. TENANT'S PASTURES

5.1 O.S. No. 346. 7.81 ha	—4/5 year ley undersown Barley 1987
5.1 O.S. No. 347. 10.04 ha	—3-year ley direct seeded autumn 1989
5.3 O.S. No. 349. 8.56 ha	—Permanent ley direct seeded autumn 1988
5.4 O.S. No. 350. 7.20 ha	—Permanent ley sown 1976

6. SOD VALUES on O.S. Nos 412 and 418—10.8 ha

7. UNEXHAUSTED MANURIAL VALUES
UNEXHAUSTED VALUE OF LIME APPLIED
RESIDUAL VALUE OF FEEDING STUFFS CONSUMED

The unexhausted value of fertiliser and lime applied to the land and the residual value of purchased feeding stuffs and home grown corn consumed on the holding.

8. TENANT'S IMPROVEMENTS

8.1—Four bay steel and asbestos dutch barn 90′ × 30′¼ erected in 1976.
8.2—Tile drains laid in O.S. No. 350 (serving 2.5 ha) carried out in 1980.

9. TENANT'S FIXTURES (Taken over and paid for by the Incoming Tenant)

9.1—Sheep dip, foot bath, four pens with 7-rail tanalised post and rail fencing, together with all concreting, water service and drains. Constructed in 1984.

9.2—Post and pig netting fencing with two strands of barbed wire on tanalised post and rail fencing—410 m run dividing O.S. No 886. Erected in 1986.

SUMMARY

1. Produce	£12,078.50
2. Growing crops	4,621.40
3. Cultivations	450.00
4. Labour to Farmyard manure	271.20
5. Tenant's Pastures	2,420.00
6. Sod values	387.00
7. Unexhausted manurial values, Unexhausted lime and residual value of feeding stuffs consumed	1,281.25
8. Tenant's improvements	2,480.00
9. Tenant's fixtures	870.00
	£24,859.35
Less an allowance for DILAPIDATIONS as shown on attached ANNEX*	9,241.00
Net Valuation	£15,618.35

Net Valuation
Dated: this 20th day of February 1990

We, the undersigned, having inspected the above, DO VALUE, the same in the sums given.

.... F.R.I.C.S. F.A.A.V.
Chartered Surveyor,
Acting on behalf of the Outgoing
Tenant

.... F.S.V.A. F.A.A.V.
Incorporated Valuer
Acting on behalf of the Incoming Tenant

*NOTE A list of the dilapidations is normally attached to the Inventory. This shows the items of claim excluding the figures agreed per item—merely a total is given at the end.

This list is not attached to this particular example inventory, but would more less take the form of the items of claim (excepting any items deleted on negotiations, given in the example schedule of Dilapidations in Chapter 11 prox). A total of the agreed claim is given.

CHAPTER NINETEEN
Compulsory Purchase Claims

19.1 The post war years have seen a dramatic increase in claims arising on compulsory purchase, largely on land being acquired for housing, schools, sewerage works, motorways, roads, road improvements etc. The law is governed by:

19.1.1. Land Compensation Act 1961

Section 5 lays down six rules for assessing compensation as follows:

Rule 1 No allowance is to be made because the acquisition is compulsory

Rule 2 The value is to be the open market value assuming a willing seller

Rule 3 Special suitability for a statutory purpose or where there is no market apart from the special needs of a particular purchaser is to be disregarded

Rule 4 Increase in value for uses contrary to the law is to be disregarded

Rule 5 Special and rare cases only can be dealt with on a cost of reinstatement basis (generally where it is used for a purpose for which there is no market e.g. Churches)

Rule 6 Rule 2 above not to affect assessment of compensation for disturbance on other matter not based on the value of the land

19.1.2. Compulsory Purchase Act 1965

Compensation for injurious affection and severance—S.7.
Compensation for yearly tenants or tenants for less than a year—S.20 (see also the 1973 Act)

19.1.3. Land Compensation Act 1973

Claims for road noise and other 'physical factors' (e.g. Smell, Lighting etc.) where no land is taken from the Claimant—Part I.

Severance of an Agricultural Holding—Sections 53 to 62
Covering expenses of removal, home loss payments, re-housing etc., and in particular farm loss payment—S.34, 35 & 36
Planning blight of Agricultural Unit—Sections 79–81
Tenant's security of tenure—S.48

19.1.4. Agriculture (Miscellaneous Provisions) Act 1968 S.12—Entitlement to Re-organisation Payment

See S.48(5) of 1973 Act. Compensation of four times the rent (or apportionment). This additional amount is tax free, but has to be deducted from the value of the tenancy interest (see 18.3.(a)). Valuers sometimes fail to appreciate that it is the rent actually payable and not rental value.

19.2. OWNER/OCCUPIER CLAIMS

These claims broadly cover:

19.2.1. Open Market Value of the land acquired at the time of the claim being settled or the date possession is taken, whichever is the earlier

A claimant is entitled in assessing the value of the acquired holding to consider splitting the property in a prudent fashion and valuing in units, if this produces a higher total market value. Such an approach may, however, affect the claim for severance or injurious affection. The possibility of development value should be considered and where appropriate an application for a Certificate of Alternative Development made. Where granted, this can add considerably to the claim. (Section 17 of the 1961 Act as amended by the Local Government Planning & Land Act 1980.) In certain circumstances, the Acquiring Authority may be able to make a set off for betterment to other land of the claimants, arising directly from the Scheme e.g. where the L.P.A. would grant permission for a petrol filling station, but only in consequence of the road improvement scheme.

The treatment of milk quota can be a very important matter where part of a dairy farm is acquired. In the vast majority of cases, the claimants (owner, occupier, landlord or tenant) will agree that the entirety of the quota will remain on the holding. If this was not the case an apportionment is made in accordance with the *Puncknowle Farms Ltd* v. *Kane* (1985) decision. The M.M.B. must be notified of the transfer and appropriate procedure must be followed.

19.2.2. Severance and Injurious Affection

In addition, claims for severance and injurious affection may well arise e.g. a farm split by road and injury to the value of the farm as a unit or parts of it including proximity of a new road to the dwelling house, interference with sporting rights, depreciation arising on loss of land resultant in certain fixed equipment being superfluous to the needs of that holding e.g. 30 acres of a 170 acres farm acquired. A specialist milking set up can no longer be used to optimum capacity. Also, additional costs of cultivations, spraying, harvesting etc., which arise where parts of fields are acquired. Depreciation in the value of the claimants retained land by virtue of added fencing liability comes under this heading (*see Cuthbert* v. *Secretary of State for the Environment 1977*).

Care needs to be taken to ensure that losses for severance and injurious affection are assessed on the basis of the depreciation in the market value of the claimants retained interest. Arguably the market will take little account of a farm with over capacity in the milking parlour but these losses will be reflected in a before and after approach. The Lands Tribunal have rejected claims on a number of occasions where these have been based upon capitalised future costs.

19.2.3. Disturbance

This generally forms a major part of a road acquisition and some other similar cases. However, these losses must be a direct result of the acquisition. Negligence by the Contractor should normally lead to a claim against him and not the acquiring authority. See, however, 19.5 below. Possible items of claims include:

(i) Loss of crops, cultivations, pasture, U.M.V.'s, R.V.F.S., R.V. Lime
(ii) Loss of profits on a crop almost ready for harvesting (intrinsic profitability is reflected in the value of the land)
(iii) Expenses incurred in moving or moving stock

(iv) Damage to crops e.g. dust settled on a silage crop
(v) Damage resultant in rock being blasted onto adjacent fields
(vi) Cost of rounding up stock which has strayed because the fencing provided was not stockproof
(vii) Injury caused to stock as a result of the works e.g. wire left about
(viii) Where herbage seeds are grown, the loss arising where certain fields or part fields, alongside the new road, can no longer be certified (and thus used for such production) owing to pollination by grasses growing on the new road verges
(ix) Cost of extra cleaning of dwellings including carpets and curtains arising from dust emanating from the works in progress
(x) Flood damage arising as a result of the works
(xi) A neighbours bull gets into a field (owing to defective road fencing) containing young beef heifers and serves them
(xii) Damage to drains
(xiii) Interference with a shoot on a temporary basis (note a claim for permanent damage will arise under injurious affection)
(xiv) Loss on forced sale of stock
(xv) Claimants time properly spent with agents, contractors, engineers etc. or indeed any time spent by the claimant mitigating his loss
(xvi) Claims for redundancy payments arising where employees are made redundant consequent on the acquisition
(xvii) Cost of re-painting and decoration occasioned by dust arising from the works
(xviii) Loss where the works interfered with a farm shop or similar retail business
(xix) Expenses properly incurred including legal and surveyors' fees in searching for and purchasing an alternative property (where the whole is acquired)
(xx) Severance of water supplies
(xxi) Blasting causing broken windows and fractures to walls and buildings

All the above matters have been encountered by valuers acting on behalf of claimants, particularly in the case of acquisitions for new roads. It is impossible to foresee the extent of the claim for disturbance when completing the claim in reply to a Notice to Treat but it is advisable that a reservation should be made that this additional claim will follow and will be quantified when the work is complete. The golden rule is to take a Record of Condition prior to entry, preferably agreed with the Authority, together with good photographs and also to advise the claimant to keep a detailed diary of every incident arising and the time he has spent on the matter. The total hours used are added up at the end of the work and a suitable claim made for the loss.

19.2.4. Farm Loss Payment

This may arise where the whole of an agricultural unit is acquired. This is payable in addition to the compensation detailed in 19.2. 1–3 above. The claimant must be a freeholder (or hold a lease with 3 years unexpired) and must, no later than 3 years after his dispossession, begin to farm elsewhere. Section 35 of the 1973 Act deals with the method of calculating the claim. The claim is complicated but briefly is the mean of three years annual profit after deducting a rental allowance.

Note: A claim is disqualified if the claimant *leaves in advance* of his dispossession or if the acquisition is made under the Blight provisions.

19.2.5. Home Loss Payment

This arises when a home is lost by compulsory acquisition, after being occupied by a claimant for at least 5 years. The amount of the payment is (under Section 30 of the 1973 Act) ten times the rateable value, subject to a maximum of £1,500 and a minimum of £1,200. It is expected that now domestic rates have been abolished, a flat rate will be fixed. The claim has to be lodged within six months of dispossession.

19.2.6. Accommodation Works

These are works carried out on the claimants' remaining land in partial mitigation of the claim for injurious affection. In strictness there is no right to accommodation works but it is normal to negotiate accommodation works as part of the compensation and these may include the following:

(a) Replacement of fences, replanting of hedges and maintaining these until established and protecting the same. [The loss in market value arising from future burden of maintenance should be reflected in the injurious affection of the land remaining. The Department of Environment will accept liability for maintenance of motorway fences but no others. It is therefore important, as fences perish and hedges need maintenance, protecting and laying, fully to reflect the loss and future liability. It has been held that the loss falls under Rule 2 (market value loss) and not disturbance as mentioned in Rule 6 and this item of claim should be included under injurious affection. (*Cuthbert* v. *Secretary of State for the Environment 1977*).
(b) Provision of gates. It is considered new gates should always be at least 4 m wide, galvanised and the gateways stoned.
(c) Removal of hedgerows where justified where odd areas of fields are left and these can be let into adjoining fields but this will mitigate the claim for severance.
(d) Provision of water supplies, troughs and meters (and pay the proper loss resulting from any additional meters installed in consequence of the acquisition) where water supplies are severed. Where supply pipes have to pass under a new road, it is wise to have these in ducts or sleeves so that such pipes can be withdrawn and replaced if ever considered necessary. It is important to note that these ducts or sleeves should be of ample size to withdraw and insert pipes. Cases have arisen in the early days of motorway construction of such ducts etc., being too small, the new road has since settled, the ducts severed and it is now impossible to replace any services in need of replacement.
(e) Continuance of severed land drains, including provision, where appropriate, of a header drain to discharge into the roadside drains (full rights for such discharge may be reserved in the conveyance).
(f) Replanting of shelter belts, specimen trees, ornamental trees etc., on adjacent land retained. This will also cover mainten-

ance until established. In practice, the Acquiring Authority may pay the claimant to do the work himself. (It may also be possible to arrange for works to be done on the land acquired, but this is by agreement and not strictly speaking, accommodation works.)

(g) Noise insulation and double glazing, special ventilation and, if necessary, venetian blinds.

(h) Provision of access to severed land (partly by means of land acquired from neighbours) by the provision of over bridges or under-passes. Note, that in the case of the latter, it is important that they should have ample headroom to take high loads and allow tractors with high cabs to pass and the former should be of adequate width. Also, many under-passes are sunk into the ground and problems have arisen with these getting flooded in wet weather. Such a bridge or pass, will also involve consideration of re-sorting existing farm roads to co-ordinate with the crossing point. The main structure of the overbridge or underpass is not an accommodation as it is constructed on acquired land and the claimant will need to reserve an easement of way with liability for future maintenance passing to the acquiring authority. Access aprons and ramps, however, are constructed on the claimants' land and properly constitute accommodation works.

(i) Making good damage to part structure acquired e.g. part buildings taken.

Note: The claimant does not have an entitlement to a 'like for like' replacement and if agreement cannot be reached the Acquiring Authority will pay compensation rather than carrying out the accommodation works.

19.2.7. Fees

Fees in accordance with Ryde's Scale (1984) should be claimed, plus travelling and other allowable expenses of Valuers and V.A.T. V.A.T. will only be payable if the claimant is not registered for V.A.T. but the majority of farmers will be registered. It is very important that the valuer should be keep a detailed record of his visits and expenses. *Fees should also be claimed on the value of accommodation works negotiated.* Also proper legal costs. Legal costs are payable by Statue and need not be the subject of Claim but it is as well to mention them.

19.2.8. Interest and Advance Payments

This is payable at the Statutory Rate on the unpaid balance of the settled claim. Advance payments of compensation (90% of the District Valuer's estimate and payable when entry is made) may be available under Section 52 of the 1973 Act. Continuing applications for further advances should be made if the District Valuer increases his estimate of the proper compensation. Advance payments of interest (which will avoid the problem of interest being paid all in one tax year) are likely to be introduced.

19.2.9. Plan

It may be a great nuisance to a landowner to have say 4,500 sq m of land acquired from say 3 fields. He does not know the individual area of the remainder. The Acquiring Authority should therefore be requested to prepare and supply the claimant with a plan of the area of the farm adjacent to the acquisition, where part O.S. numbers have been acquired. This plan should show the individual area remaining in acres and hectares of each affected enclosure.

There is no obligation on the Acquiring Authority to provide plans but in suitable cases the costs in preparing such plans would form a claim under Rule 6.

19.3. TENANT'S CLAIMS.

Tenant's claims are governed by the same Acts as previously mentioned and reference is made, in particular to the decision given in the cases of *Wakerley v. St Edmundsbury B.C. 1977 & 1979* and *Anderson v. Moray D.C.* The matter is very complex but very simply the tenant is entitled under Section 20 1965 Act to:

(a) the value of the unexpired term of the tenants interest (in his tenancy and ignoring any prospect of a Notice to Quit founded on the Authority's scheme)
(b) the payment that would be made to him by an incoming tenant e.g. value of growing crops, U.M.V.'s, R.V.F. Stuffs,

unexpended value of lime applied, tenant's pastures, tenant's improvements etc.

(c) loss or injury sustained, such as disturbance, loss on forced sale of stock, removal expenses, home loss payment (if living at the house continuously for ·5 years—see the 1973 Act (Note: Payments under Section 34(2) of the 1948 Act have no place in a claim under Section 20)

(d) severance where this can be proved and injurious affection to the remainder of the farm in that tenancy (separate tenancies do not qualify)

Any contractual bar on assignment is to be ignored because this is overridden by the concept of a notional sale, Rule 2 (19.1.1)

In order to avoid duplication of claims, Section 48(5) of the 1973 Act provides that the above compensation is subject to a deduction of the payment of four times the annual rent paid (or the apportionment) in respect of the land taken (Section 12 of the 1968 Act). [For basis of apportionment see 19.6.4.] This of course applies where the claim for the interest is in excess of four times the rent.

Care should be taken to note whether the dispossession is by *Notice to Quit* (under the power of resumption of possession in the tenancy agreement, if it exists) i.e. where the Authority has already bought the freehold, or by *Notice of Entry*. (see S.11 1965 Act).

Where entry is made by Notice to Quit the tenant is not entitled to be paid for the value of his interest (a) above [but see below] being merely entitled under the Agricultural Holdings Act 1986, Section 60 to a minimum of one years rent and a maximum of two years but he will, in addition, get the four years rent re-organisation payment under Section 12 of the 1968 Act. However, if it is to his advantage he may claim to go out under the compulsory purchase code, rather than under his tenancy agreement in which case a Notice must be served under Section 59(2) 1973 Act. In this event the claim is made under Section 20, as above, but should not then include an item for Section 60 Agricultural Holdings Act 1986.

Where entry is made by Notice of Entry, the claim is assessed under Section 20 of the Compulsory Purchase Act 1965 and comprises all of items (a), (b), (c) & (d) above, subject to the deduction under Section 48(5) of the 1973 Act. The Section 12 (1968 Act) payment of four times the rent is thus preserved as a tax free payment but the deduction is made to prevent double counting.

19.4. VALUATION OF THE TENANT'S UNEXPIRED INTEREST

This is a very difficult valuation to estimate especially in the case of a strip of land taken from a farm. If a whole farm is acquired it will prove easier to assess as evidence can be found of compensation paid by Landlords to their tenants for vacating and there are cases known where the compensation has ranged from merely off-setting dilapidations against tenant right to over £1,000 per acre being paid for possession being given. The following factors need considering:

(a) the age of the tenant. If he is a young man say in his thirties, and has no desire to vacate, he is most unlikely to consider vacating unless very substantial compensation is paid. His chances of obtaining another rented farm are very remote and he will probably lose his living. If, on the other hand, the tenant is near or over retiring age and has no family hoping to succeed to the tenancy, the value of the interest is substantially diminished, especially if his health is not good (but see (f) below). This is particularly important where the method of valuation is founded on the surrender value to the Landlord

Most Acquiring Authorities would argue that it is necessary to assess market value in accordance with Rule 2 and hence the necessity to envisage a willing seller. Since the 1977 Act (now case G 86A) a landlord has been unable to serve an uncontestable notice to quit on the death of the original tenant, where the tenancy has been assigned. In these circumstances the age of the tenant has assumed less importance.

(b) the possibility of a member of the tenant's family being able to apply for tenancy succession. Although such a family member has no legal interest whatever in any claim, this is bound to be a factor that would influence the attitude of the tenant and it is contended that it has a definite bearing on the value of the tenant's interest.

(c) how important any given area of land is to a particular tenant? If only a smallholder with the farming being of a high intensive standard, *any* area of land acquired would be far more important to him than in the case of the tenant of a large holding which is not intensively farmed. In other words, the latter might be more likely to accept a relatively smaller compensation for vacating a given area of land than the former.

(d) whether there is any possibility (other than in the case of the acquisition) of the landlord being able to obtain planning consent for *any other user* than agriculture and consequently obtain possession of the area of land involved. [Note: The Act requires the Valuer to leave out of account the prospect of a notice to quit founded on the actual scheme underlying the acquisition—Section 48 1973 Act.]

(e) is the tenant enjoying a profit rental? If on a long lease with *no* rental revision provisions he may be, but in other cases, the maximum period of profit rental he could claim would be three years.

The extent of the tenants improvements and the value to be disregarded under Schedule 2 2 (1)(a) of the 1986 Act is relevant.

(f) the possibility of assignment (even if prevented by the tenancy agreement). It is necessary to consider a hypothetical assignment or surrender to the landlord i.e. the law presumes a willing seller, which in turn means leaving out of account any bar on assignment).

To summarise, the tenant is compensated *for his interest as if it existed in a no-scheme world*, being at risk of dispossession solely in so far as his tenancy agreement and the statutes would allow i.e. the Valuer must ignore any possibility of a notice to quit founded on the Authority's scheme.

Several different approaches have been taken by valuers to assess the value of the *tenants interest.* Lands Tribunal case of significance is the Wakerley Case (*Wakerley* v. *St. Edmundsbury Borough Council* 1977 which went to the Court of Appeal in 1979). This, however is not a particularly helpful case and may not have another parallel. The case of *Anderson* v. *Moray D.C.* (1980) decided that since the land in question had development potential, an effective notice to quit could be served on the tenant and this had the effect of reducing his security of tenure. Amongst the suggested approaches made by valuers are:

(a) A proportion of the difference between the vacant possession value and the value of the reversion
(b) A proportion of the vacant possession value of the area acquired
(c) A proportion of the difference between the vacant possession value of the freehold interest and its value subject to the life tenancy. This will have regard to the age of the tenant
(d) Capitalising the profit rent for the remainder of the tenant's lifetime
(e) Capitalising profits for a reasonable period at a rate which reflects risk taking.

From the resulting valuation there is to be deducted the 4 × rent apportionment re-organisation payment (provided the sum calculated exceeds this sum). This, however, is then claimed separately as a payment which is free both of income and capital taxes.

It has been stated that valuation is not an exact science. This statement is no where more true than when valuing an agricultural tenant's unexpired interest. In many cases there will be considerable variance between the claimant and his valuer's views and that of the District Valuer and often the claimant will be dissatisfied with the final compensation agreed or awarded. Until tested in the market, there is never a 'definite' value, but rather a range of values between which the answer may be reasonably be expected to be.

19.5 PROCEDURAL POINTS

From experience it is desirable for a Record of Condition of the land to be prepared and agreed in road acquisitions, prior to entry. Photographs should be taken. These can be very useful at a later date.

It is considered that to start with, no claims should be made against contractors. In the first instance the matter should be referred to the resident engineer but all claims should initially be lodged against the acquiring authority who are ultimately responsible.

Contractors frequently wish to hire sites for soil dumps, spoil removed, working space, depots etc. They are usually prepared to pay what might appear to be a good consideration for such user. These sites should always be licensed and not let, a proper agreement entered into and full payment made prior to entry. Contractors should be responsible for fencing and full reinstatement. When using land for depots, the long term damage caused can be considerable since contractors usually lay hardcore, erect hut bases, lay concrete and form roads. It should be stipulated that the top soil should be removed and stored prior to any work being done. All debris should be removed on completion, including all hardcore, stones, rock etc., the land subsoiled at least three times, top soil replaced, fertilised and re-seeded. The consideration charged should reflect that the land affected is unlikely to be fully productive for several years. Hence the inference above 'might appear to be good consideration' is not necessarily so. The acquiring authority is not concerned with these private arrangements.

19.6. EXAMPLE CLAIMS

19.6.1. Owner/Occupiers Claim

Moorend Court is an 85 ha (210 acres) farm about $\frac{3}{4}$ mile from a market town in the south-west. It has a period 5 bedroomed Georgian residence and extensive, principally modern buildings (many of which have been erected in the last 10 years). It also has two semi-detached farm worker's cottages. 110 Pedigree Fresian Cows are kept, together with about 40 followers. There is also a bull beef unit producing about 50/60 bulls a year. About 50 acres of cereals are grown.The land is all Grade 2 and the soil is free-draining loam derived from old red sandstone. The farm has a primary wholesale milk quota of 556,000 litres.

The Department of Transport are about to construct a single carriageway by-pass which affects Moorend Court as shown on the plan. All legal formalities have been complied with and the Compulsory Purchase Order in respect of the scheme was confirmed on 21st December 1986. Entry was taken on 1st April 1987.

The By-Pass will severely affect Moorend Court severing a large part of the land from the homestead. The new road will be within 110 m of the residence.

At various meetings held between the County Engineer (acting on behalf of the Department of Transport), the Resident Engineer, the District Valuer, the Claimant and his Agent, it has been agreed:

(a) an access bridge will be built across the By-Pass to give access to the homestead from road B.007
(b) a concrete farm road will be built from points 'A' to 'B'
(c) the existing water main supply will be re-laid in a 300 mm diameter sleeve under the by-pass between points 'C'–'D'
(d) a noise bund will be constructed on the new boundary opposite the residence
(e) all drains affected will be connected the surface water drains to be constructed alongside the new carriageway
(f) a two row staggered quickthorn hedge is to be planted on the new boundary and is to be protected on the field sides with a pig netting fence on tantalised posts with two strands of barbed wire
(g) the milk quota will be retained by the claimant (*Puncknowle* case).

The area of land acquired is 6.07 ha (15 acres). 33 ha (82 acres) of the farm will be severed from the homestead. Licences are also being taken to re-align the brook in O.S. No. 7 and bordering O.S. Nos. 5 & 8.

The duration of the contract was to be 1 year from the date of entry. However, for a variety of reasons, the contractors were on the land until 15th July 1989. The over-bridge took much longer to complete than anticipated and there were problems of access to the severed land with resultant delays in fertiliser application, silage making and inadequate access to get the dairy herd across to the set stocked grazing pastures. The temporary fencing erected was not stockproof and was frequently damaged by the contractors and stock got out. Further damage was caused when entry was taken to remove this and plant the quickthorn hedges. Damage was also caused when the contractors entered to locate a blocked drain which caused flooding. There were major problems with dust resultant from earth moving in 1987. This dust affected the residence, gardens, all buildings, pastures and silage cuts. The spring of 1988 was exceedingly wet and very muddy conditions existed until 1st June. The Water supply to both the homestead and the severed land was cut and milk yields lost. Rock blasting resulted in a bill of £542 for damages to a precision chop silage harvester and silage making was delayed for two days. 1 ha ($2\frac{1}{2}$ acres) of grazing was lost owing to inadequate fencing for 10 months.

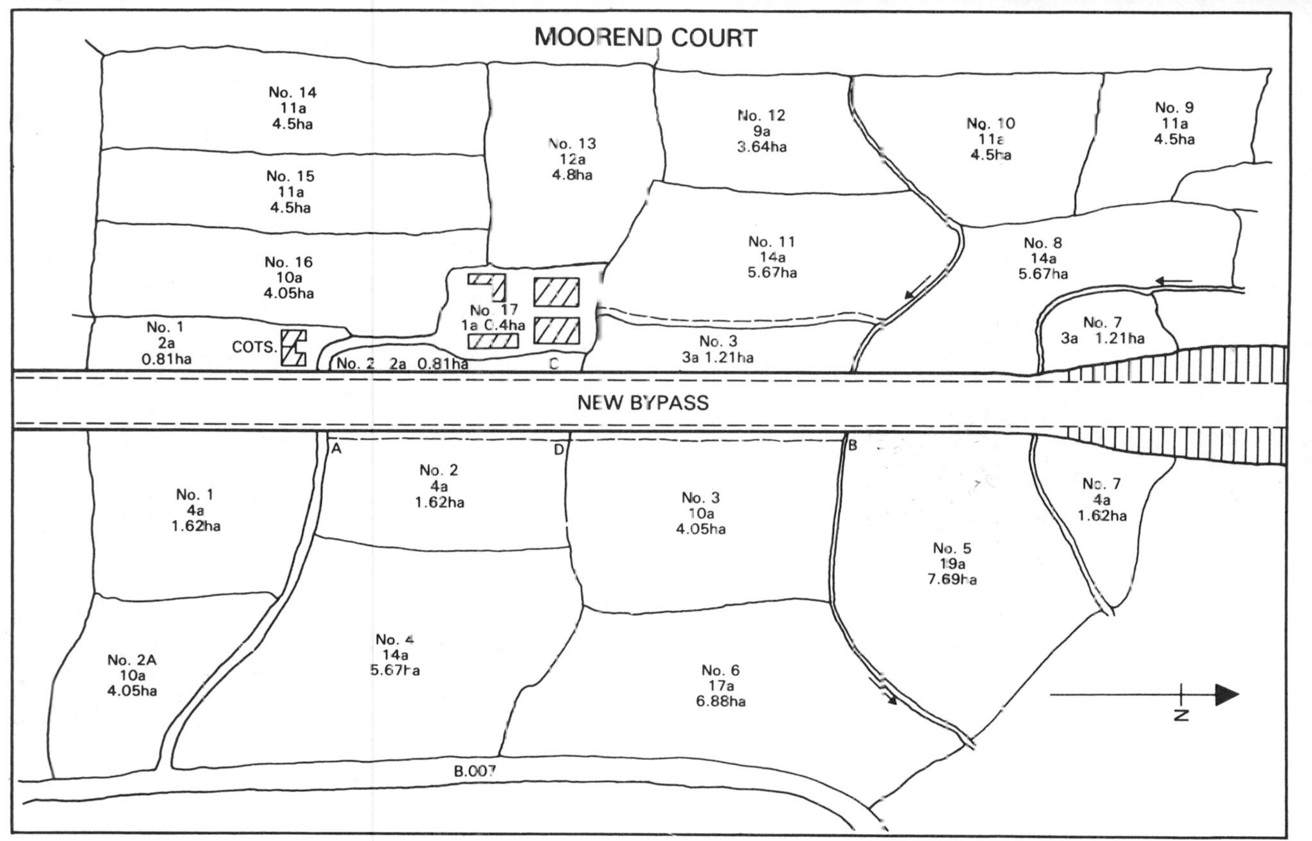
MOOREND COURT
No. 14 11a 4.5ha
No. 13 12a 4.8ha
No. 12 9a 3.64ha
No. 10 11a 4.5ha
No. 9 11a 4.5ha
No. 15 11a 4.5ha
No. 11 14a 5.67ha
No. 8 14a 5.67ha
No. 16 10a 4.05ha
No. 17 1a 0.4ha
No. 7 3a 1.21ha
No. 1 2a 0.81ha
COTS.
No. 2 2a 0.81ha
C
No. 3 3a 1.21ha
NEW BYPASS
A
D
B
No. 2 4a 1.62ha
No. 7 4a 1.62ha
No. 1 4a 1.62ha
No. 3 10a 4.05ha
No. 5 19a 7.69ha
No. 2A 10a 4.05ha
No. 4 14a 5.67ha
No. 6 17a 6.88ha
N
B.007

19.6.2

STATEMENT OF CLAIM

by

Mr. J. R. Yeoman (Freeholder & Occupier)

against

The Department of Transport

following the acquisition of part of

MOOREND COURT, WESTINGFIELD, ANYCOUNTY

for the construction of the Westingfield By-Pass,

April 1987.

Prepared by: WILLIAM DAVIES & REES
Agricultural Valuers &
Chartered Surveyors,
Dolanog,
Welshpool,
Powys.

Pt. O. S. Nos. 1, 2, 3, 5, & 7—Area 6.07 ha (15 Acres)

1. LAND TAKEN

1.1 Freehold Interest Acquired.

The Acquiring Authority to pay for the freehold interest acquired—6.07 ha @ £8,125 ha. £49,318.75

1.2 Licence

The Acquiring Authority to pay for the user of the area of .6 ha used as a licence for the duration of the contract i.e. 2 growing seasons. The sum of £1,482.00

Total for Land Taken. £50,800.75

2. SEVERANCE

Area severed east of the By-Pass is a total of 33 ha (82 acres). Cow grazing pastures have previously been O.S. Nos. 2, 3, 11, 5, 7, & 8. It will no longer be practicable to use O.S. 5 & 7 (east) and O.S. 9 will now have to be used instead. There will also be additional travelling time for men and equipment to collect the cattle from O.S. No. 9 and for husbandry in the severed land. Cattle will also be travelling twice a day for part of the grazing season into O.S. Nos. 5 & 7 (east) and will have to be cleaned up at wet times apart from the damaging effect of cow dung on tarmacadam and also general wear and tear. Additional estimated time taken due to the By-Pass severance is 150 hours per annum. The Acquiring Authority to pay loss in value of 33 ha due to severance at £675 ha.

Total Severance £22,275.00

3. INJURIOUS AFFECTION

3.1. The residence is severely affected by traffic noise, fumes, car lights shining at night, dust and its view is severely impaired. Current value of residence as part of the whole farm is estimated at £150,000. Loss 30%	£45,000	
3.2. The pair of cottages are similarly affected. Current value of each is £50,000. Loss 30%	£30,000	
3.3. The efficient running of the holding is also severely affected by the proximity of the By-Pass. Certain areas cannot, in future, be used for herbage seed production and in all approximately 12 ha will be affected by noise, fumes, dust etc. There is also the risk of stock getting out onto the By-Pass. The occupation of the farm buildings are also effected. Loss:		
3.3.1. 12 ha capital value of £8,125 per ha depreciating by 10%	£9,750	
3.3.2. Buildings with capital value of £150,000 depreciating by 10%	£15,000	
3.3.3. Buildings and equipment with a capital value of £15,000 (£12,500 buildings; £2,500 Equipment) will become redundant. Loss	£15,000	
3.3.4. 1020 m of fencing has to be maintained. This will need annual maintenance in perpetuity and the hedge, when of sufficient height (with about 10 years growth) will need laying.		
Depreciation in value in the remaining 79 ha (195 acres) remaining in the claimants ownership @ £200 per ha	£15,800	
Total Injurious Affection.		£130,550

4. TIMBER & AMENITY

Loss of 10 mature oak trees and 2 specimen copper beech trees in O.S. 3 (west)		
4.1. 30 cu m oak and beech @ £60	£1,800	
4.2. Loss of amenity and shelter 12 trees @ £50	£600	
Total claim for Timber		£2,400
Total Amount of Claim		£206,025.75

5. DISTURBANCE & CROP LOSSES ETC

This is to be the subject of an additional claim to be made when the contract is completed.

6. ACCOMMODATION WORKS

The Acquiring Authority to execute at their expense and to the claimants complete satisfaction, the following accommodation works.

6.1 Plant a staggered two row quickthorn hedge on the new boundary protected on the field side by pig netting and two strands of barbed wire on tanalised posts and maintain, including annual weeding, until fully established.

6.2 Grub out and level the hedgerow between O.S. Nos. 7 (west) and 8; O.S. 4 & 2; 11 & 3 (west) and 7 & 5 (east).

6.3 Remove the redundant farm road in O.S. No. 11 and fill the roadway with top soil.

6.4 Connect all exposed drains to the new roadside drains to be laid. Also to repair all drains affected by the works and to allow, at all times, any new drains laid by the claimant (or his successors in title) on the holding to discharge into the roadside drains.

6.5 Supply, fix and connect to the main water supply water trough to serve O.S. Nos. 3 & 5. This will entail laying a main from the existing farm supply and providing a P.V.C. sleeve to take the main under the road—300 mm diameter encased in 150 mm concrete.

6.6 The By-Pass will restrict the existing irrigation facilities to the land to the east. It will be necessary to provide No. 2 300 mm diameter sleeves at points to be agreed, under the road to take the irrigation main.

6.7 Double glaze the windows of the farm residence and cottages and provide any further noise insulation considered necessary.

6.8 Erect a suitable screen to minimise noise and traffic lights for a distance of 60 m run in front of the house.

7. PLAN

The Acquiring Authority to prepare a new farm plan and calculate the area of the fields affected and supply No. 5 copies to the claimant.

8. ADVANCE PAYMENT & INTEREST

An advance payment of 90% of the claim to be made as soon as entry taken.

9. FEES

The Acquiring Authority to pay the Ryde Scale contribution of claimant's valuers and solicitors fees. Advance fees to be paid.

10. INTEREST

Interest at the Statutory Rate to be paid on the remaining amount of the claim (after payment of the advance) from the date of entry to date of payment. Such interest to be paid on an annual basis.

E & O. E.

PRACTICE NOTES

(a) *Claim Item No. 8 Advance Payment and Interest*

The advance payment will be made on the basis of the District Valuers estimate of the compensation payable and not upon the amount of the claim.

(b) *Claim Item No. 9 Fees*

In view of the volume of work involved in claims of this nature, few valuers are prepared to accept the Rydes Scale as being their fee.

19.6.3.

STATEMENT OF CLAIM FOR DISTURBANCE

by

Mr. J. R. Yeoman (Freeholder & Occupier)

against

The Department of Transport

as a result of acquisition of part of an entry onto

MOOREND COURT, WESTINGFIELD, ANYCOUNTY

to construct the Westingfield By-Pass

August 1989

Prepared by: William Davies & Rees,
Agricultural Valuers &
Chartered Surveyors,
Dolanog,
Welshpool,
Powys.

The Acquiring Authority to pay for the following matters:

1. The Electricity supply was cut on 14th March 1988, for 18 hours. One milking lost and another delayed and 2 cows affected by mastitis.

   Loss of 1241 litres milk @ 18 p. £223.38

   Loss on two cows £300 (one lost a quater) and vets bill £70. £370.00
   2 mens overtime—6 hours @ £6. £36.00

2. Access was denied to the homestead on 15 separate occasions.
   Inconvenience and Loss @ £20. £300.00

3. Cost of additional cleaning of the farmhouse and two cottages due to dust and mud over a period of 50 weeks.
   1 Hour per day. 350 hours @ £2.50. £875.00

4. Cost of additional cleaning of carpets and curtains (receipts attached). £68.20

5. The farmhouse and two cottages require re-decorating externally and internally one year earlier than usual due to the works. £750.00

   Cost £3,000 over 4 years. $\frac{1}{4}$ cost.

6. Access drive was left in such a bad state that farm lorry axle broke. Cost of repairs (receipt attached). £268.75

7. Additional cost of cleaning farm drive arising from the work.
   40 hours @ £3. £120.00

8. Water supply to O.S. Nos. 3, 5 & 7 (east) were cut on July 5/6 1988 and 1620 litres of milk lost @ 18 p. £291.60

9. Water supply fracture resulted in 20,000 gallons of water being lost.
   Cost @ 150 p per 1000 gallon. £30.00

10. Dust polluted the grazing in O.S. Nos. 3, 5 & 7 for the period of 10th May–28th July (less 2 weeks) 1987. Damage to 25 acres. Milk yields suffered and it is considered that yields were depressed by 8,000 litres @ 18p. £1,440.00

11. Dust polluted 2 cuts of silage in 1987, (total 280 tonnes). It became necessary to feed, on Ministry of Agriculture, Fisheries and Food recommendation (copy available) additional concentrate and 10 tonnes 16% feeding stuffs were purchased from West Central Farmers Limited at a cost of £135 per tonne. £1,350.00

    Quantity of silage totally useless.
    Loss 40 tonnes @ £15. £600.00

12.	Precision chop silage harvester damaged by rock blasted. Cost of repairs (copy account attached).		£542.00
13.	Blasting has caused fracture in south elevation of farmhouse. Estimated cost of repairs		£1,500.00
14.	Ditches silted up in O.S. Nos. 3/5 and 5/7. Cost of clearing out 200 m @ £1.50		£300.00
15.	Part O.S. Nos. 11, 7 (west) and 8 flooded due to blockage of stream by contractors. 3½ acres affected. Loss of grazing £50 and silage crop £120		£170.00
16.	Damage caused by Contractors entering O.S. No. 7 (east) in June 1988 to locate and rectify blocked drain.		£82.00
17.	Damage caused to all enclosures affected when contractors entered during very wet conditions in November/December 1988 to plant quickthorn hedge and remedy defective post and wire protective fencing.		£140.00
18.	Loss of Crops—1987.		
	Loss of 1 acre Avalon Winter Wheat in O.S. No. 1.		
	3 tonnes @ £140 & Straw £20	£440	
	Additional Cost of Harvesting	25	£465.00
19.	Stock got out on six occasions viz: 15th, 16th & 18th April; 20th June, 21st & 28th July 1987. Cost of rounding up 12 hours @ £3.		£36.00
20.	Damage was caused to the silage crop in O.S. Nos. 8 & 7 (W) on 15th, 16th and 18th April 1988. Loss 3 tonnes @ £20		£60.00
21.	To claimants time wasted arising from the entry. A diary has been kept and this shows 41 hours spent by the claimant attending meetings with his solicitors, valuers, N.F.U., resident engineers, district valuer and considerable time telephoning largely to report complaints.		
	41 hours @ £12	£492	
	56 telephone calls @ £1	56	
	20 letters written @ £2	40	
	161 miles travelling @ 30p	£48.30	£636.30
22.	Forced Sale of Stock		
	The loss of the land and general upheaval caused will result in 20 cows having to be sold off, many at the wrong time for their optimum sale i.e. as stale cows. A capital loss arises on this forced sale 20 cows @ £200		£4,000.00
23.	Valuers Fees & Costs		

24. Other Matters Arising

E & O. E.

19.6.4. Tenants Claim

Assume all facts as given above except that the holding is occupied by Mr. John Farmer on an annual Candlemas (2nd February) tenancy at a rental of £12,500 per annum (revised last on 2nd February 1987). The tenant, who is 45 years of age and has two sons of 18 & 16 years of age, holds under a normal agricultural tenancy and repairing liabilities are in accordance with S.I. 1473. He has improved the farmbuildings since he took the tenancy in 1974 and has provided most of the modern buildings (these are tenants improvements executed with unconditional landlords consent) and all the dairy equipment including the 6/12 herringbone parlour. Because he has carried out these improvements he is enjoying a profit rental of about £35 per ha. The holding has a current investment value of £3,250 per ha and a vacant possession value of £8,125 per ha.

The first action to take in formulating this claim is to agree the rental apportionment between the Tenant, landlord and the Acquiring Authority. This is necessary because the law requires it (S.19 Compulsory Purchase Act 1965) and also to enable the tenant to be paid the Statutory Payment ('additional amount') by virtue of the 1968 Act.

Basis of Apportionment

This is not a rent 'reduction' under S.33 1986 Act.The apportionment of the rent payable at the date of severance should be on the basis of the value attributable to the severed land and the retained land, respectively, as part of the whole. The rent apportioned to the retained land should have no regard to any depreciation suffered by reason of the severance or the use made of the severed part.

The apportionment of the rent dies not operate to create a new tenancy and therefore does not debar the landlord from securing an increase (or the tenant a decrease) on the occasion of the next normal review date in accordance with Section 12 of the Agricultural Holdings Act 1986. The effect of any severance will then be reflected in the rental review, and any advantage or disadvantage to landlord and tenant, in the interim, should be reflected in the valuations for the respective compensation claims.

The four heads of claim are set out in 19.3; everything to be claimed must therefore be fitted into one or other of the heads.

STATEMENT OF CLAIM

by

Mr. J. FARMER (TENANT)

against the Department of Transport
following the acquisition of part of
MOOREND COURT, WESTINGFIELD, ANYCOUNTY

for the construction of the Westingfield By-Pass

Prepared by: Williams Davies & Rees
Agricultural Valuers &
Chartered Surveyors,
Dolanog,
Welshpool,
Powys.

Area Acquired (6.07 ha) forming Part of O.S. No. 1, 2, 3, 5, 7 & 8

The acquiring authority to pay the following claims:

1. VALUE OF TENANTS UNEXPIRED TERM

1.1 Value of Tenants Interest in area taken

The tenant is a young man and can reasonably expected to be farming the holding for a further 20 years or so when one of his sons can, provided he satisfies legal requirements, be expected to succeed to the tenancy. (Agricultural Holdings Act 1986, Part IV, Sections 34–58).

Furthermore he has carried out major improvements to the holding. Normally the value of the tenants interest is considered to be 50% of the difference between the estimated vacant possession value (£8,125 per ha) and the Investment Value (£3,250 per ha). However, in view of the tenants improvements in this particular case, it is considered that the value here is 60% of the aforementioned difference

= £4,875 × 60% (6 ha) = £17,550			
Less: 4 × £882 (apportioned rent) =	£3,528	£14,022.00	
(see also Item 4 of claim)			

1.2. Licence Area

User of the area of .6 ha as a licence for the duration of the contract i.e. 2 growing seasons		£1,482.00	
Total Value of Tenancy interest			£15,504.00

2. OUTGOING TENANTS PAYMENT

2.1. Unexhausted Manurial Values	£91.00	
2.2 Residual value of Feeding Stuffs consumed	£55.60	
2.3 Unexhausted value of lime applied (Detailed calculation of 2.1—2.3 attached)	£62.20	
2.4 Tenants pastures in Pt. O.S. Nos. 3, 5 & 7–4 ha @ 120	£480.00	
2.5 Loss of crops in O.S. Nos. 1, 7 & 8–2 ha @ £250	£500.00	
2.6 Labour to Farmyard Manure in O.S. Nos. 3 & 5	£210.00	
Total of Tenant Right		£1,398.80

3. INJURIOUS AFFECTION TO TENANCY

3.1 Fencing Liability

Injurious affection arises as a result of 1020 m of additional fencing having to be maintained. This will need annual maintenance and when the hedge is of sufficient size (approx. 10 years growth) it will need laying.

Additional annual expenses inc:

(a) 80 hours maintenance @ £4	£320		
Y.P. 4%	£25	£8,000.00	
(b) Cost of laying hedge in 10 years time 1020 m @ £6	£6,120		
Defer 10 years @ 10%	.38	£2,325.00	
3.2 Redundant buildings and equipment. Current value of redundant buildings is £80,000–20% redundancy		£16,000.00	
Total Injurious Affection			£26,325.00

4. AGRICULTURE (MISCELLANEOUS PROVISIONS) ACT 1968—1 Sec. 12

Claim for re-organisation payment £882 × 4 = £3,528.00

5. DISTURBANCE, CROP LOSSES ETC.

This is to be the subject of an additional claim to be made when the contract is completed [as for the owner/occupier]

Note: Four times the rent of the affected area will be deducted from the claim, (provided the agreed amount of claim is higher than the four times rent; as shown in 1.1, see also Item 4).

6. ACCOMMODATION WORKS

As item 6 of Owner/Occupier Claim.

7. REDUCTION IN RENTAL

A reduced rental of the holding is consequence upon the acquisition to be negotiated and agreed between the Claimant, the Landlord and the Acquiring Authority.

8. PLAN

The Acquiring Authority to prepare a new farm plan and calculate the area of the fields affected and supply No. 5 copies to the claimant.

9. ADVANCE PAYMENT

An advance payment of 90% of the claim to be made as soon as entry taken.

10. FEES

The Acquiring Authority to pay the claimant's Valuer and Solicitors fees

11. INTEREST

Interest at the Statutory Rate to be paid on the remaining amount of the claim (after advances) from the date of entry to date of payment. Such interest to be paid on an annual basis.

E & O. E.

19.7. CLAIM BY LANDLORD i.e. THE REVERSIONARY INTEREST

The same general principles of compensation apply to land let as apply to owner/occupied land but it should be noted that under the provisions of Section 48 of the 1973 Act, the Landlords' interest to be valued is that which actually existed at the date of Notice to Treat, *ignoring the effect on the tenants security of tenure on the Acquiring Authorities proposals or any other compulsory purchase scheme.* Any Notice to Quit served by the landlord, founded upon the Authority's scheme, is to be ignored and if the tenant has already left it is to be assumed that he has not done so. The landlord can also make a claim for injurious affection and severance, but it is unlikely in practice that there would properly be any claim for disturbance items, other than surveyor's fees.

Practise Notes (applicable to all claims)

(1) It is the duty of the claimants agent to submit a full and detailed claim.
(2) The District Valuer is unlikely to agree to all terms of claim or the amounts claimed. Nevertheless, it is up to the claimant and his agent to substantiate the claims made.
(3) Frivolous items of claim should not be lodged.

(4) Each case must be treated individually. Varying matters of claim will arise for different claims and the valuer should not miss items which obviously are valid.

(5) Where tenancies exist accommodation works should be agreed between the Landlord, Tenant and the Acquiring Authority. Similarly any rental reductions consequent upon the requisition, but the Acquiring Authorties are not too concerned on this matter and are reluctant to become involved.

(6) In appropriate cases the tenant may be able to substantiate a claim for severance although this is usually reflected in the reduction of rent that may be negotiated.

CHAPTER TWENTY

Farm Rents

20.1. Provisions for varying farm rents are contained in Section 12 & Schedule 2 of the Agricultural Holdings Act 1986.

20.2. The basic requirements of a Notice requiring Arbitration to vary a rental is that it has to be served to take effect from the next termination date following the demand for arbitration. On an annual agricultural tenancy this means that a demand for arbitration must be served at least one year prior to the term date. Rents can only be varied every three years. However, it should be noted that some holdings, let on lease, have their own rent revision provisions included and these agreed provisions must be observed especially as some do not necessitate a Notice requiring arbitration on the rental to be even served.

20.3. Schedule 2 of the 1986 Act provides that the *rental properly payable in respect of a holding shall be the rent at which the holding might reasonably be expected to be let by a prudent and willing landlord to a prudent and willing tenant*, taking into account all relevant factors, including, (in every case)

(a) the terms of the tenancy (including those related to rent)
(b) the character and situation of the holding including the locality in which it is situated.
(c) the productive capacity of the holding and its related earning capacity and
(d) the current level of rents for comparable lettings as determined in accordance with subsection (5)

20.4. *Productive Capacity* means what can be produced from the holding (taking into account fixed equipment and any other available facilities on the holding e.g. very good or very poor buildings, irrigation facilities etc.) on the assumption that it is in the occupation of a competent tenant who is practising a system of farming suitable for the holding.

20.5. *Related Earning Capacity* means that the extent to which in the light of the productive capacity, a competent tenant practising such a farming system could reasonably be expected to profit from farming the holding.

20.6. Paragraph 1(3) of Schedule 2 provides that the Arbitrator shall take into account evidence of the current level of rents payable or likely to become payable for comparable lettings. This evidence may be cases where rents are fixed by agreement or by arbitration under this Act or are likely to become payable (when tendered) in respect of tenancies of comparable agricultural holdings on terms (other than terms fixing the rent payable) similar to those of the tenancy under consideration. He shall, however, disregard

(a) the scarcity element i.e. the overbid to secure a tenancy
(b) any element of those rents, which is due to the fact that the tenant, or a person tendering a rent for a comparable holding, is in occupation of other land in the vicinity so that they may be conveniently occupied together.
(c) any effect on those rents due to any allowances or reductions made in consideration of charging premiums.

20.7. Paragraph 2(1) of Schedule 2 further instructs the Arbitrator to disregard any increase in the rental value of the holding due to

(a) tenants improvements or fixed equipment other than those provided by virtue of an obligation imposed on the tenant by the terms of his contract of tenancy.
(b) Landlords improvements in so far as he has received government grants in respect of the execution of these improvements.

20.8. Paragraph (3) of Schedule 2 further states that the Arbitrator

(a) shall disregard any effect on the rent on the fact that the tenant who is party to the arbitration is in occupation of the holding and
(b) shall not fix a rent at a lower amount by reason of any dilapi-

dation, deterioration or damage to buildings and land caused by the tenant.

20.9. The effect of Section 12 of the Act is that there are now three very important concepts in farm rental valuations. The Arbitrator must take into account

(a) the productive capacity of the holding and its related earning capacity
(b) evidence of the current level of rents for comparable lettings or even rents being tendered.
(c) the arbitrator must ignore the scarcity element in a rental tender.

20.10. THE PRODUCTIVE CAPACITY OF THE HOLDING

This means basically what a holding is capable of producing in the occupation of a competent tenant practising a system of farming suitable to the holding. Various types of farms exist and some differing types are as follows:

(a) Diary Farms
(b) Grazing or Feeding Farms
(c) Arable Farms—suitable only perhaps for cereal and oil seed rape production
(d) Good class Arable Farms with deep, free draining, stone free, soil, capable of growing potatoes, sugar beet and similar high value cash crops
(e) Rearing Farms—generally marginal land holdings
(f) Hill Farms
(g) Specialist Farms e.g. Pig and Poultry Holdings

There is, therefore, a wide spectrum of holdings all which have differing productive capacity. Climate, elevation, aspect, location etc., all affect the productive capacity as do the quality of soil. The valuer should, however, not place too much reliance on the Agricultural Land Classification Maps. They are of some help but they do not always signify the real quality of the soil concerned e.g. most riverside land is shown as Grade III, largely because it is liable to flooding. However, most riverside land is good quality freedraining soil which generally speaking has far greater productive capacity than say some Grade II land and sometimes even Grade I land. Again the productive capacity of fields even on the same farm varies. Some fields have extremely good quality soil which may be capable of producing, in an average season well over 3 tonnes of cereals per acre ($7\frac{1}{2}$ ton per ha) whilst others on the same farm may only produce, on average, say 2 tonnes (5 tonnes per ha). The gross margin difference here on say a winter wheat crop may be as much as £120 per acre (£300 per ha). Other areas of a farm, or even a whole farm may be capable of being irrigated, which despite being somewhat costly, can boost yields considerably in drought conditions and thus increase the productive capacity. A contrast is the case of say a 300 acre (120 ha) Cotswold arable farm with rather shallow stony brash, limestone soil compared with a similar size Herefordshire farm, in the Wye Valley with some alluvial silty soil and some free draining red sandstone soil. Assume both have equal quality buildings including adequate corn storage and good general purpose buildings. The Herefordshire farm, however, has both potato and beet quotas whereas these crops could not be satisfactorily grown on the Cotswold farm.

The Herefordshire holding could also, by feeding beet tops and other roots grown, finish off a substantial number of store lambs and thus take advantage of the high prices realisable for them, fat in March when they are sold off.

The differing productive capacity of these two holdings can be illustrated as follows:

120 ha Herefordshire Farm

Crop/Ent.	*ha*	*Yield per ha (T)*	*GM*	*Total GM*
Winter Wheat	25	7.4	630	15,750
Winter Barley	25	6.3	510	12,750
Sugar Beet	16	44	949	15,184
Potatoes	8	44	1,752	14,016
200 Breeding Ewes	38		31.40	6,280
60 Fattening Steers			50	3,000
500 Root Lambs			7.66	3,830
Spring Barley	8	4.6	380	3,040
	120			£73,850 £615.42 per ha

120 ha Cotswold Farm

Crop/Ent.	*ha*	*Yield per ha (T)*	*GM*	*Total GM*
Winter Wheat	25	6.2	514	12,850
Winter Barley	50	5.4	431	21,550
400 Breeding Ewes }	45		31.40	12,560
60 Fattening Steers }			50	3,000
	120			£49,960 £416.33 per ha

It will, therefore, be seen, that on two similar sized holdings, with substantially the same system of farming, but one with better and more versatile land, capable of producing higher value cash crops, the total G.M. is better on one by almost £24,000.

20.11. Illustrating further the factors that influence the productive capacity of any holding, reference is made to Chapter 17 with most of the matters referred to therein, being relevant. However, to summarise these, the following matters should all be considered by the Arbitrator and Valuer in determining the productive capacity of any given holding:

(*a*) *Land and Soils*
Is it level or is it banky?
Does the land flood?
Are there irrigation facilities?
Is the land versatile i.e. suitable for tillage and grazing.
What is the rainfall?
Elevation and aspect?
Is the land well watered, shaded and fenced?
Is the soil freedraining?
Is the soil deep?
Is the soil stony?
Is it early soil or is it cold and late?
Are there wet areas in need of drainage?
Is the soil heavy or is it light?
Can high value drops be grown *and harvested*?
Is there a high water table?
What is the M.A.F.F. Grading of the soil?
Can one winter stock on the land without undue poaching?

Are there any waste areas?

(*b*) *Buildings*
What is the standard of buildings and are they suitable and adequate for the type of farming system practised e.g. is there good grain storage and drying facilities on a cereal farm and is the dairy set up suitable and adequate on a dairy farm?
Are the buildings modern and versatile e.g. contrast the existence of a large general purpose building (suitable for almost any use) and the presence of several smaller older type buildings of inadequate size and with too low a headroom.
Are the buildings too specialised in nature and thus restrictive in their possible user?
Are the buildings well designed and of good access?
Are repairing costs likely to be average?

(*c*) *General*
Position of the farm—is it convenient or isolated from towns and markets?
What quality of house exists?
Are there sufficient cottages?
Does the farm have quotas? (It is clear that quotas, whether they are potato, beet or milk have great value and can have considerable effect on the earning capacity of any given holding and thus its rental and capital value.)

20.12. RELATED EARNING CAPACITY

The earning capacity, as the wording of the section states is, of course, related to the productive capacity of the holding, in fact, it directly follows. The definition in the Section states that it is the 'extent to which, in the light of that productive capacity, a competent tenant practising a system of farming suitable for the holding, can reasonably be expected to profit from farming the holding'.

This has the partial effect of limiting a rental an Arbitrator can award to a reasonable share of the surplus an efficient farmer, carrying on a suitable farming system, can earn. A farmers accounts may here be of value to a valuer but the accounts for a single year only, can be misleading, for a variety of reasons. If accounts are to be produced, it is considered, that to be fair to both landlord and tenant, they should be for three consecutive years, if possible. It should be remembered that sometimes but not always, accounts are produced for tax purposes and do not always reveal the correct profitability.

Forward budgets can also be of immense value and it might be argued, that if produced for one year only, (when prices are known) they may be more valuable than past accounts, which are historic. Both accounts and a forward budget should be produced as they are essential in the rental calculation.

What is a fair share of the Net Farm Income (total Gross Margin less total Fixed Costs) that should be paid as rental? Opinions vary—some land agents and valuers consider that it should be 60%. Others, considering that capital a tenant farmer employs (particularly in livestock enterprises) and the somewhat poor returns on the investment, think that 40% is reasonable. This is a matter of contention, but the average percentage that most reasonable minded valuers adopt is 50%, but experience shows that it varies from 40—55%.

Any budget prepared for a rental calculation should be prepared with great care and should provide as much detail as possible as otherwise it could be worthless.

20.13. CURRENT LEVEL OF RENTS PAID FOR COMPARABLE LETTINGS

Paragraph 1 of Schedule 2 requires the Arbitrator to take into account any available evidence of rents payable for comparable lettings. This evidence can take the form of any agreed rents, arbitration awards, or rents which are likely to be tendered in respect of comparable holdings. However, the Arbitrator must disregard any element in such rents which is a 'scarcity element'. Most valuers will however, concede, that when advising applicants for farm tenancies of the amount to tender, a sum is invariably added to a calculated rent to make such tender sufficiently attractive for serious consideration of the tender by the landlord or his agent. Such 'overbid' or scarcity element in rental valuations must be disregarded.

The Arbitrator must also disregard any element of rent due to the fact that the tenant (or tenderer for land), of any comparable holding is in the occupation of other land in the vicinity, that may be conveniently occupied together with that holding. This simply means that the Arbitrator disregards the case of a farmer paying perhaps a very high rent for a nearby area of land (and who can well afford to pay over the odds) since he is effectively spreading his fixed costs over a larger acreage e.g. perhaps he can farm such additional holding with the same labour and equipment and thus incur very little additional fixed costs and can afford to pay an excessively high rent.

Evidence of comparable rents, or the general level of rents paid can be very useful to an Arbitrator and Valuer. However, farms are not like a row of semi-detached houses and direct comparison of farms is well nigh impossible. Quality of land, fixed equipment, approach, situation, elevation, topography etc. etc. dictate that there are rarely direct comparable holdings.

If comparisons are cited, the farm concerned should always be inspected, the tenancy agreement perused, and all necessary information on tenants improvements etc. obtained. In practice, evidence of comparable lettings must be looked into very thoroughly if it is to be of value.

It will also be worthwhile for an Arbitrator and Valuer to consider the annual average rental statistics published by the Ministry of Agriculture. However, these must be treated as a guide only but can be very helpful if only to indicate general trends.

Furthermore, any effect on rents due to allowance or reductions made in consideration of the charging of premiums, must be disregarded.

20.14. TENANTS IMPROVEMENTS

Paragraph 2 of schedule 2 states that the Arbitrator shall disregard any increase in rental value of the holding due to tenants improvements or fixed equipment (other than that provided by him as an obligation in his contract of tenancy). *It is immaterial whether landlords consent for such improvements was received or not* and the farm must be valued as if the improvements or equipment did not exist. In practice, this is difficult and valuers often value the property as it actually is and then value out such improvements. However, this method should be approached with care as the end result may otherwise well result in over-emphasing the value of the improvements, as part of the whole farm. Alternatively the farm may be valued as unimproved i.e. the improvements are ignored. Examples of cases of valuing out are:

(i) 10 acres of very rough grazing reclaimed by the tenant. Previously it had little value (say £3 per acre) but is now good land worth, on the open market, say £35 rent per acre. The calculation is:

Current rental value	£350
Value before improvement	30
Increased net rental value	£320

The net cost of the improvement was £2,500 and provided it is property farmed, in the future, the improvement will be beneficial to the occupier 'ad infinitum'. The increased rental value, charging interest at 12% on £2,500 = £300. This expenditure has therefore released an extra or hidden value (latent value) of £20 per annum (£320 − £300 = £20).

The County Court decision in *Tummon* v. *Barclays Bank Limited* (1979) decided that 'latent value' should belong to the landlord. Logically this is because the landlord owns the raw material for the improvement—here the land (and this could also be argued for the site of a tenants buildings or other improvement). The landlord should therefore be given credit for the basis on which a tenant derives benefit. In this example the tenant is not entitled to deduct more than £300 per annum from the gross rental viz:

Increased Rental value	£320
Less: Latent Value	20
Increased net rental value	£300

(ii) Erection of 3,600 sq ft (335 m^3) covered yard 10 years ago. Total life expectancy is 40 years. Current net costs of the work including site levelling, would be £21,500. Unexpired life is 30 years. Current value is

$$£21,500 \times \frac{30}{40} = £16,125 \text{ say } £16,000$$

However, the latent value of the site is say £50 per annum (in view of the use now put to the site it is possible that even a higher latent value can be properly justified).

A landlord erecting such a building for his tenant would reasonably expect 12% interest on his investment. A tenant could also expect a similar return. The current net increased rental value resultant on the improvement is therefore:

£16,000 × 12% = £1,920 − £50 (annual latent value)
= £1,870 which sum should therefore be deducted from the gross rental.

(iii) The tenant has divided a 30 acre field into two convenient sized fields by the erection of a 400 m length of pig and barbed wire fencing, together with a 3.6 m iron gate. The work was carried out three years ago and the current cost of the work would be £710. Total expectation of life is 12 years. The annual value of the improvement is therefore:

$$£710 \times \frac{9}{12} \text{years} \times 12\% = £63.90 \text{ (say } £64)$$

It is considered, here that no latent value has been released and the sum thus deductable from the gross rental value is £64 per annum.

20.15. RENTAL VALUATION

In preparing a rental valuation under Section 12 of the 1986 Act, the following course of action is suggested:

(i) A detailed inspection of the holding should be made noting carefully:
   (a) type of fixed equipment, quality of land (which the valuer should carefully assess, grading this himself, into say four categories of differing qualities and carefully assessing its productive capacity and thus its earning capacity having regard to all relevant factors).
   (b) tenants improvements and fixtures, if any, and calculating their annual value.
   (c) how well farmed is the holding and is it being 'competently' farmed?
   (d) is the system of farming practised suitable for the holding.

(ii) Prepare a detailed budget, based on the farming system

practised and also taking into account whether the tenant is farming to optimum levels of production on a system suitable for the holding.

(iii) Examine, if possible farming accounts for the past 3 years. However, these may not be available or will not be produced.

(iv) Consider all available evidence of rentals paid in the area and possibly, if the farms are comparable, further afield.

(v) Research into the terms of the tenancy. Nowadays many new lettings are on a full repairing and insuring basis which means that any tenant holding on these terms bears a higher burden than a tenant holding on reparing terms equating to those prescribed by S.I. 1473 of 1973. It is difficult to generalise on the additional burden a tenant takes on a full repairing and insuring basis, but this is likely to add an extra 15% to the norm.

Valuers have differing methods of valuation which often produce not dis-similar results. All the factors previously mentioned above and in Chapter 17 should be taken into account. However, at the end of the day (but taking into account the Statutory requirements of Section 17 of the 1986 Act) what is most important of all is the quality of the soil, its versatility, quality of the fixed equipment and that it is competently farmed to a system suitable for the holding concerned.

20.16. EXAMPLE VALUATION

HEN BLAS, PENGWERN, MONMOUTH, is a 210 acre stock rearing/dairy farm in the County of Gwent with a small modern 3 bed residence, a semi-detached 3 bed workmans cottage and a reasonable set of farm buildings. The land is generally undulating with some severe banks and 95 acres is Grade III and 103 acres of Grade IV land on the Agricultural Land Classification Map. About 14 acres are rough unproductive land and the homestead and access roads. The land lies between 400′ and 700′ above sea level. The valuer has placed the land in his own categories as follows:

A—94¼ Acres B—57¾ Acres C—44 Acres D—14 Acres.
The farm is not well situated being 17 miles from the nearest market

town and is approached for the last mile by a very narrow council maintained road.

The tenant, aged 44 is very competent and better than average farmer. He runs the farm with the aid of a 24 year old workman and some casual labour at busy seasons. 70 milking cows are kept (milk quota is 350,000 litres and it has been informally agreed that 90% belongs to the landlord and 10% to the tenant). Some calves are reared, 15 being dairy replacements and 30 others as beef stores (some Friesian Steers, some Friesian Heifers and the remainder Hereford cross Friesian Cattle) are sold off as 15 month stores. 110 full mouthed Suffolk cross Ewes are kept. In view of the type of holding this is, production is considered to be at optimum level and without intensification and capital investment, cannot be further increased.

The tenant has carried out the following improvements at his own expense:

(a) concreted 210 sq m around the buildings over the last 6 years. Net cost £1,560
(b) erected cow kennels for 60 head
(c) erected milking parlour, constructed collecting and dispensing area, together with feeding area (inc. provision of mangers). Erected dairy
Note: (b) and (c) was done 3 years ago at a total net cost of £26,000. Some old buildings belonging to the landlord were demolished on site.
(d) tile drained 21 acres (8.5 ha) 5 years ago at a net cost of £3,850
(e) provided a piped water supply with No. 4 concrete water troughs to provide a mains water supply to 56 acres (22.5 ha) of the top land. Net cost was £1,210 and the work was done 2 years ago.

The farm is held on an annual Candlemas Tenancy with the tenant being responsible for all repairs and payment of insurance premiums. Current rental fixed by Arbitration with effect from 2nd February 1987 is £5,000 per annum.

The best approach to this rental valuation is threefold:

(a) prepare a detailed budget as shown. It is suggested that in this case 45% of the Net Farm Income (Gross Margin less

Fixed Costs) should be adopted as a reasonable guide to the rental payable. This budget should reflect the productivity and related earning capacity of the holding.

(b) prepare a detailed breakdown rental valuation of the holding. This valuation should carefully reflect the quality and value of the house, buildings and land. The valuation should be on a full market rental basis showing the separate gross values of:

the residence,
cottages (if any),
buildings,
the four categories of land (or less if there are lesser categories).

This will produce the gross rental value from which must be deducted the calculated annual rental value of the improvements. It is also suggested that if the holding suffers from some severe limiting factors (which have not been taken into account in the gross rental value computation) such as poor approaches, isolation, an awkwardly shaped farm and other constraints, percentage deductions of between 1 & 10% should also be deducted from the gross value as should any undue repairing and insuring burdens. If, on the other hand, the farm has some outstanding features, not already accounted for, such as good position, irrigation facilities etc., there should be added a suitable percentage to the Gross Value to reflect these. Also any other special advantages such as the inclusion of sporting rights in the tenancy should be valued.

The result should produce a fair and balanced rental valuation.

(c) consider any evidence of comparable lettings. The exact terms of the tenancy should be ascertained and suitable adjustments made if anomalies exist between these and the subject land. Also adjustments made to reflect quality of land, scarcity element etc.

(A)	FARM BUDGET APPROACH		
	All prices based on Current Values		
(a)	Land		
	Farmable Land		198 Acres (80.16 ha)
	Rough, Buildings etc.		12 Acres (4.84 ha)
	Total		210 Acres (85.00 ha)
(b)	Enterprises		
	Dairying		
	70 Dairy cows @ 1.4/Acre/Head (.56/ha) Head		98 ac. (39.6 ha)
	15 Dairy Replacements @2.5/Acre/Head (1/ha) Head		37 ac. (15.0 ha)
	(1 Unit = 1 Calf, 1 yearling & 1 Heifer)		
	30 Beef Cattle @ 1.2/Acre/Head (.48ha/head)		36 ac. (14.5 ha)
	110 Ewes @ 4/Head/Acre (.1 ha/head)		27 ac. (10.9 ha)
	Total		198 ac.
(c)	Gross Margin Summary		
(1)	70 Dairy Cows @ £583 Head	£40,801	
(2)	15 Dairy Replacement Units @ £291/Unit	4,365	
(3)	24 Head Store Cattle @ £90/Head	2,160	
(4)	110 Ewes @ £32.80/Head	3,608	
	Total G.M.		£50,934
(d)	Fixed Costs		
(1)	Labour		
	1 full time man (inc. overtime)	£7,250	
	Casual	1,000	
(2)	Machinery Costs inc:		
	Depreciation, repairs fuel & oil, contract, sundries £60/Ac. (£150/ha)	11,880	
(3)	Water and Electricity	1,350	
(4)	Miscellaneous inc:		
	telephone rates insurance, general maintenance, sundries	£6,100	
	Total Fixed Costs		£27,580

(e) Net Farm Income

(G.M. less Fixed Costs)	
Gross Margin	£50,934
Less Fixed Costs	27,580
Net Farm Income	£23,354
50% of N.F.I. as Rent =	11,677
Deduct for Tenants Improvements say 25%	£2,919
Net Rent Payable	£8,758

The remainder of the N.F.I. represents the return to the tenant on his labour, management and investment income.

GROSS MARGIN ANALYSIS

(1) Dairy Cows

(70 head—Quota 350,000 litres = 5,000/Cow)

Output		
5,000 litres @ 18p	£900	
Calf (allowing for losses)	115	
	£1,015	
Less: Cow Depreciation	60	£955
Variable Costs		
Concentrates £1.48 T @ £150	£222	
Vet & Medicines	20	
A. I. & Recording	20	
Sundries & Bedding	25	
Forage Costs	85	£372
Gross Margin per Cow		£583

(2) Dairy Replacements

Each dairy replacement unit represents 1 Calf, 1 Yearling and 1 Heifer.

Output		
Value of down calving heifer	£650	
Less Calf value	110	
Net output		£540
Variable Costs		
Concentrates inc. calf food	£140	
Vet & Medicines	17	
Bedding	20	
Forage Costs	72	£249
Gross Margin per head		£291

(3) Store Cattle

30 head are kept and selling
at 15 month age = 24 no. annually

Output		
Sale of 335 kg. Steers @ 120p/kg	£402	
Less: Cost of Calf	£130	
Net Output		£272
Variable Costs		
Concentrates	£110	
Vet & Medicines	12	
Misc. Bedding, Forage etc.	60	£182
Gross Margin/Head		£90

(4) Sheep

Lambing % 1.5

Output		
$1\frac{1}{2}$ lambs × 18 kg @215 p/kg	£58.05	
Wool	3.00	
Ewe Premium	7.00	
	£68.05	
Less: Ewe & Ram Dep.	£11.40	£56.65
Variable Costs		
Concentrates	£ 8.00	
Vet & Medicines	3.85	
Sundries & Forage	12.00	£23.85
Gross Margin/Ewe		£32.80

(B) RENTAL VALUATION APPROACH

$94\frac{1}{4}$ acres Class 'A' land @ £40		£3,770.00
$57\frac{3}{4}$ acres Class 'B' land @ £32		1,848.00
44 acres Class 'C' land @ £26		1,144.00
14 acres Class 'D' land viz:		
12 acres Rough @ £5	60.00	
House @ £35 p.w.	1,820.00	
Cottage @ £22 p.w.	1,144.00	
Buildings @ £60 p.w.	3,640.00	£6,664.00
Gross Rental		£13,426.00

210 acres Total

Less Deductions

			fwd. £13,426
(1)	Tenants Improvements		
(a)	Concreting Depreciation over 20 yrs. Work done 5 yrs. Cost (net) £1,560		
	Current value £1,560 × $\frac{15}{20}$ = £1,170 @ 12% =	£140	
	No latent value		
(b)	Cow Kennels, Milking Parlour, Dairy, Feeding Area etc. Cost £26,000. Work done 3 yrs. Latent value in old buildings say £500 p.a. 10 year depreciation		
	Current value £26,000 × $\frac{7}{10}$ = £18,200 @ 12% = 2,184		
	Less latent value 500	£1,684	
(c)	Drainage Cost £3,850 5 yrs. ago 20 yr. depreciation Latent value say £210 p.a. (£10 per acre)		
	Current Value = £3,850 × $\frac{15}{20}$ = £2,887 @ 12% = 346		
	Less latent value 210	£136	
(d)	Piped water supply Cost £1,210 2 yrs. ago. 10 yr. depreciation No latent value.		
	Current value = £1,210 × $\frac{8}{10}$ = £968 @ 12% =	£116	£2,076
(2)	Landlords Share of Repairs and Insurance		
	(a) Repairs	£2,000	
	(b) Insurance	500	£2,500
(3)	Disability of Holding Isolated and a poor shape with considerable uphill pull to the homestead, which is badly placed in relation to the land and with banky land.		
	Deduct 3% of gross	402	
			£4,978
	RENTAL VALUE		£8,448

(C) COMPARABLE RENTALS

Two comparisons on the same estate, in the locality, are available as follows:

FARM 1

A 120 acre (48.5 ha) farm, of similar land, but with much better buildings (all the landlords) and used purely as a dairy farm. House is good. Tenant is holding on a full repairing and insuring term. Milk quota is 420,0001. The farm was let by tender on a Candlemas Tenancy, 2 years ago. There was considerable interest with 61 parties inspecting, 38 made tenders ranging from £40 to £70 per acre (£99–£173/ha). The farm was eventually let to a young man at £50 per acre (£123.50/ha).

FARM 2

A 320 acre (130 ha) stock rearing farm 3 miles away. This has a good house and traditional buildings, except that it has a modern sheep shed. The farm has hill rights for 100 ewes. The tenant is in his mid fifties and is a good farmer. He has held the tenancy for 26 years and repairing liabilities are in accordance with S.I. 1473. He keeps 75 head of cattle and 350 breeding ewes and lambs. All stock is sold off as stores. The tenant has not carried out any improvements except reclaiming 21 acres of rough banks which was heavily grant aided.

Current rent fixed by agreement from Candlemas 1990 is £7500 per annum £23.43/Acre) (£57.89/ha).

Dealing with these comparisons the £57.89 Arbitrator/Valuer must inspect the farms, look at the tenancy agreements, consider all terms of letting and evaluate any tenants improvements. Having done this he lists his findings as follows:

FARM 1

(a) New tender letting which in his opinion reflects a scarcity element of approximately 20% in rental.
(b) A smaller farm which does not use any hired labour except perhaps casual.
(c) Has a good milk quota.
(d) Has much better buildings.
(e) Tenant has not carried out any improvements.
(f) Better situated farm.
(g) No waste land.

FARM 2

(a) This is an old established tenancy.
(b) Repairing liabilities are different from subject farm.
(c) This is a completely different type of holding that is not strictly comparable, particularly as it is a purely stock rearing/hill type farm.
(d) Tenant has done some improvements in reclaiming 21 acres but this area had quite a valuable latent value.
(e) Farm has the benefit of hill subsidies.

His conclusions in these two cases are:

FARM 1—This farm is to a degree comparable to the subject farm but rental paid reflects scarcity value. Deducting this and also making the other necessary adjustments, the rental of the subject farm is indicated to be in the bracket of £40–43/Acre (£100–£106/ha).

FARM 2—This farm is not comparable and is of little help except to the extent that the rent of the subject farm should be appreciably higher than in this case which was only recently fixed at just over £23.43/Acre (£57.89/ha).

SUMMARY OF CALCULATION & EVIDENCE

The result of the calculated valuation and budget and rental comparison evidence is as follows:

(a) Farm Budget (reflecting the estimated productive capacity of the holding and its related earning capacity).
Calculated Rental £8,758

(b) Rental Valuation (reflecting type of holding, terms of tenancy, tenants improvements, improvements etc.).
Calculated Rental £8,448

(c) Comparable Rental Evidence
Suggests adjusted rental should be
£41/Ac/£101.27/ha: £8,610

The average of these is £8,605 (say £8,600) which is the rental fixed (£40.95/Acre) (£101.15/ha).

NOTE: It would normally be wise to check on published rental averages and percentage adjustments (provided by the Ministry of Agriculture).

CHAPTER TWENTY-ONE

Valuation Clauses on Sales of Farms

21.1. Differing valuation clauses are included as part of the conditions of sale of farms and estates, and their extent is largely dependent on where the property sold is situated. Agents operating in certain areas—particularly the south of England—have wide, all-embracing clauses, often covering the full extent of a valuation when a tenant quits. It is thought, by some, that it is unreasonable for a vendor to expect a purchaser to pay a full tenant right type of valuation in addition to the price for the freehold.

21.2. Care must be taken in drafting valuation clauses in order to avoid subsequent disputes. The basis of the valuation should be clearly given. It is usually prudent to stipulate that hay and straw should be valued on a market value basis. Consuming value usually has no relevance in a sale of a farm since more often than not it is the vendor vacating, and not a tenant. However, should it be necessary to value hay and straw at consuming value, reference should always be made to the consuming value prices fixed by the local branch of the C.A.A.V., as this will be a firm basis for the valuation. In no circumstances should silage be stated, to be valued on a consuming basis—no such thing as consuming value exists, at the time of writing, for silage. It is best to stipulate, where the silage clamp is unopened at the date of completion, for this to be taken over at a fixed sum. If the clamp is opened and in use, it is advisable to state that the valuation should be made on the basis prescribed by Numbered Publication 154 (as subsequently amended) of the Central Association of Agricultural Valuers. Even then, a disagreement on value is a distinct possibility.

21.3. If crops are mature at the date of completion, it is advisable that these should be valued on a crop value basis, i.e. the value of the mature crop—but to avoid dispute, whether mature or not, it is best, wherever possible, to stipulate a sum to be paid for such crop. This, of course can include enhancement value, if it is thought to be applicable.

Where crops are recently sown and where cultivations carried out, it should be clearly stipulated that these are to be taken over at cost of seeds, fertilisers, sprays, and the cultivations are to be paid for at C.A.A.V. rates or contractor's costs (where they have been carried out by a contractor).

21.4. Where young seeds are to be paid for, the sum to be paid for them should be given.

21.5. Where equipment or fixtures are to be paid for, again to avoid dispute, the sum to be paid for them should be given.

21.6. In recent years it has become noticeable that the settlement of valuations is becoming an ever-increasingly long drawn out process—possibly, in some cases, to delay payment. This is unfair to the vendor, and it is thought advisable that, unless a sum is paid on account, certified by the outgoing vendor's valuer, on completion date, the balance of the finalised amount of the valuation or any arbitration award thereof should carry interest from the date of completion to the date of payment, at say, 4% above the current Bank base rate.

21.7. A clause stating that no claim for dilapidations, deterioration of any other offset should be included. This is most important, and particularly so where a farm or land is sold which was previously occupied by a tenant.

21.8. It is important that a valuation clause should be clear, specific and not ambiguous in any way, as disputes will otherwise arise. It is not sufficient to say that the valuation should be carried out in the 'usual way'. The valuation clause should state that, where applicable, valuers are to be appointed by each party and in the event of disagreement, the matter settled by an arbitrator appointed by them or failing agreement by the President of the C.A.A.V.

21.9. A common pitfall is for the term 'tenant right' to be used in referring to items to be taken to. *Tenant right has no place in vendor and purchaser relationships* and this reference should be avoided. If, for instance, unexhausted manurial values, unexhausted lime and residual values of feeding stuffs consumed, tenant's pastures etc. are to be paid for, the valuation clause should clearly and specifically state what is to be taken to and also on what basis the valuation is to be calculated.

21.10. *A suggested specimen valuation clause* is as follows:
'The purchaser shall in addition to the purchase price pay the sum of £ for the fitted carpets in the lounge, hall and staircase of the farmhouse, the young seeds sown in O.S. No. 123 and the Desco 1,200 litre refrigerated milk tank and Alfa-Laval milking machine with Brooks 3 h.p. electric motor, and shall also take over and pay for at valuation on completion, the following:

(a) all hay and straw remaining on the holding, at completion, at market value

(alternatively)

(a) all hay and straw remaining on the holding, at completion, at consuming value as fixed by the and District Agricultural Valuer's Association).

(b) silage remaining on the holding, at completion, valued in accordance with the basis prescribed in Numbered Publication 154 as amended of the C.A.A.V.

(c) the growing crop of winter barley in O.S. No. 456 (ha) at the cost of cultivations, seeds, fertiliser and sprays, together with enhancement value of £ (per acre or £ per ha).

(d) all cultivations and acts of husbandry carried out, fertilizers and sprays applied to the unplanted arable land since the last crop was harvested.

(e) labour to farmyard manure to heap in O.S. No. 789.

All acts of husbandry to be valued at C.A.A.V. costings or contractor's costs, whichever applicable.

No claim will be made for the valuable unexhausted value of fertilisers, or lime applied or the residual manurial value of feeding stuffs consumed, and there shall be no counterclaim for dilapidations (if any), deterioration or other offset made whatever.

The valuation is to be made by two valuers, one appointed by each party and failing agreement, shall be referred to a single arbitrator, appointed by the valuers or, in the event of disagreement, by the President of the Central Association of Agricultural Valuers. Such reference will be under the provisions of the Arbitration Act 1950 as amended.

If the amount of the valuation has not been agreed by completion date, then the purchaser shall pay to the vendor's solicitors, as agents for the vendor on completion date, such sum as shall be certified by the vendors valuer, as payment on account of the valuation, the remaining balance to be paid within seven days of such valuation being determined or agreed. If full payment is not made upon completion the balance will carry interest at 4% above Bank plc's base rate from date of completion to date of payment.

As an alternative, it may be preferable, dependent on the size of the valuation, to substitute the following clause:

'If the valuation is not settled and paid on completion date, the purchaser shall pay interest at the rate of 4% above Bank plc's base rate on the sum finally determined due from date of completion to date of payment.'

21.11. PRACTICE POINTS

(i) It is advisable to keep valuations as simple as possible. It has been found that excessive over-burdening valuations frequently have a depressing affect on prices obtained for the freehold.

(ii) Do not include 'petty' items in a valuation.

(iii) Word the valuation clause very clearly, always giving the basis of valuations.

(iv) Do not refer to 'tenant right' where obviously this is out of context and has no legal standing.

(v) If unexhausted manurial values, unexpired value of lime and residual value of feeding stuffs are to be paid for, do not refer to them as 'tenant right'—they are actually 'improvements'.

(vi) If labour to farmyard manure is to be charged for, only charge for this where this is a haul of some distance from the buildings. It is unfair to charge where the heap is almost alongside or very near to the building cleaned out—surely, is it not the duty of a vendor to clean out his buildings?

CHAPTER TWENTY-TWO

Records of Condition

22.1. Many land agents and valuers have in the past dismissed Records of Condition as of little value. Nevertheless, Section 22 of the 1986 Act provides for such records to be made at the request of either party at any time, and also the tenant can request that a record of improvements and tenant's fixtures should be prepared at any time during the tenancy. If either party does not agree that a record should be made, the President of the Royal Institution of Chartered Surveyors, on request, can appoint a person to make such record. Unless there is agreement on costs (and it is usual for the person requesting the record to be responsible for costs) Section 22(3) states that the costs of the record shall be borne in equal shares by the parties.

22.2. A Record of Condition is often a very valuable document more especially to a tenant when he is vacating and there is a possibility of a claim for dilapidations arising. It is also very useful in partnership farming, where the partners agree to keep a holding farmed to the standard it was and also kept in the same condition as it was when the partnership commenced.

22.3. Nevertheless, a record is only of value if it is a good detailed document which deals specifically with all aspects of the subject property. Many records prepared in the past have been altogether too skimpy and too general. A tenant has a duty to farm well, and should he take over a run-down holding he would, in normal circumstances, be expected to put the place in order in, say, 5 years. Even so a record would be of value in this case, as it would identify all fixed equipment and its condition together with all aspects of the condition of the land including state of ditches, culverts, outfall drains, fences, hedges and also the condition of the land.

22.4. Another example of the value of a record is where a farm has old, wide, very gappy, high hedgerows, which were not stockproof at the commencement of the tenancy, or even hedgerows which though shown on the O.S. map as existing, do not in fact exist. Cases have arisen where claims have been submitted for replacing gates and the missing hedgerows, or making stockproof hedgerows which were no more than a few clumps of isolated thorn and trees. Again, any nails or staples in trees can be recorded and, thus, claims avoided. The presence of certain weeds that cannot be satisfactorily and easily eradicated e.g. blackgrass, can also be recorded.

22.5. A record should always be clear, precise concise and easily followed. A scale plan should be included with O.S. numbers and areas of each field shown. In most cases, photographs should be used and properly identified in the record.

22.6. It is customary for a record prepared by a surveyor for a landlord or tenant to be checked and possibly amended by his opposite number, and for both parties to sign the engrossed record as an agreed document. This certainly has the effect of removing any doubt as to the accuracy of the record. Two copies are always provided, and these should be attached to the tenancy agreement.

22.7. *Example:*

RECORD OF THE STATE AND CONDITION

of the holding known as
FLATMEAD PARK LAND
FLATMEAD

in the county of Gwent.

Landlord:	H. J. Scrivens Esq.
Tenant:	Flatmead Farming Partnership.
Record taken and made	26th June 1989.

NOTES:

(a) Unless otherwise stated in this record, all items are deemed to be in a reasonable and tenantable order.
(b) The plan attached should be referred to for the location of some of the references made below.
(c) Photographs in Annex 1 refer to the marginal numbers shown in the Record.

LAND

1. O.S. No. Pt. 297, Pt. 304, 305 and 306
Flatmead Park. 104.498 acres.

1.1 Description

This field forms Flatmead Park and is open level parkland (now cultivated) with No. 21 oak trees and 9 horse-chestnut trees growing. Note that the original boundaries between O.S. Pt. 297 and O.S. Pt. 304, 305 and 306 have all been removed and boundaries levelled.

1.2 Cultivation

Winter barley.
Electricity transmission line traverses in a NW–SE direction.

1.3. Weed infestation

There is a severe blackgrass infestation throughout this field and wild oats are also widespread.
There is sporadic infestation of couch, charlock, groundsel, redshank and other annual weeds throughout the field.
At the headlands there are patches of infestation of nettles, docks, fat hen, thistles, broad-leaved plantain and hogweed.
There is also a widespread infestation of all these weeds plus sterile broome under all the parkland trees.

1.4. Drainage

There are several wet areas in the field, indicating either absence of drainage or a breakdown in any drainage system existing. No drain outfalls were found in the ditches.

1.5 Boundaries

1.5.1. O.S. Pt. 297/Road—east

Low hedge trimmed last season, but now overgrown and in need of trimming. This hedge is gappy, weak and is not stockproof.
There is an infestation of cleavers, sterile broome and hogweed in this hedge.
General encroachment of thistle, docks, fat hen and nettles at headlands up to 3 m into field.
Ditch (on roadside side) overgrown and in need of cleaning out.
Single-strand barbed wire protective fence in reasonable order but No. 10 loose posts.

1.5.2. O.S. Pt. 297/Road—north east

Low hedge overgrown in places and in need of trimming. Hedge infested with weeds, particularly cleavers and nettles. There are several gaps and also six sections which, in the past, have been fenced with post and 4-rail (offcuts) fencing and which is now weak and defective.
General encroachment of fat hen, nettles, thistles and brambles up to $2\frac{1}{2}$m in places at the headlands.

Ditch (on roadside) very overgrown and filled in in places and needs cleaning out. No protective fencing.
Gateway with painted metal gate and iron posts which are rusting and the top rail is buckled. Gate does not hang properly and has no latch. Hanging post loose. Gateway overgrown. Four timber side rails—all defective.

1.5.3. O.S. Pt. 207/308

Boundary belongs away.
Severe infestation of nettles, docks, thistles, blackgrass and sterile broome up to 10 m at headlands into field.
Ditch at side is very overgrown, not draining, and in need of cleaning out.
Two-strand barbed wire protective fencing in very poor order.

1.5.4. O.S. Pt./297/Road—north west

High overgrown hedge which is encroaching approx. 2 m into the fields. Hedge is weak at base and gappy. Two large gaps adjacent to two hedgerow oak trees. Hedge in need of cutting and laying. Not stockproof.
Ditch (on roadside) overgrown and fallen in in places, and in need of cleaning out. No protective fencing.
No. 2 oak trees have barbed wire stapled into them.
Painted metal gate bent and rusty, and does not hang properly.
No. 5 timber side rails badly split.
An area of scrub woodland in NW corner overgrown.

BUILDINGS

2. O.S. No. Pt. 501—2.04 acres

Description

Pt. O.S. No. 507 contains the buildings used by the farming partnership. These buildings have been in existence for a number of years and are consequently ageing and wearing. The comprise:

2.1 Grain Store 90′ × 60′

Constructed of steel, portal framed with Big Six asbestos roof, corrugated iron side cladding, at N end, E side and part S side. Concrete floor. Timber grain walls are in very poor and damaged condition, with timber boarding generally damaged and defective, and 48′ run is missing. Corrugated iron gable end at S side damaged and rusting. Steelwork all rusting and paint peeling off. Stanchion at S.W. end damaged by impact and buckled. The asbestos roof is weathered and lichen is growing to extensive areas of the east side. Half-round guttering uneven and have no fall, and all joints leaking. Stop end missing at N.E. end. All gutters need cleaning out. Asbestos rainwater downpipe at S.W. has bottom section cracked and shoe missing.
Concrete floor uneven and wearing, with some cracks evident.

2.2 Lean-to Implement Shed (attached to Grain Store)

Five-bay open-sided steel and Big Six asbestos roofed Implement Shed with part hardcore and part earth floor. Corrugated iron gables. Steelwork rusting, as is the corrugated iron. Asbestos guttering in similar condition to Grain Store but both rainwater downpipes missing.

2.3 Fertiliser Store

Constructed of concrete block walling with asbestos roof. Single steel framed 6-pane window and ledged, braced and battened door.
Concrete floor.
Serious vertical settlement crack the length of the E wall. No 2. window panes broken and No. 1 cracked. Window rusty. Door handle missing, and door secured with hasp, staple and padlock.
Door damaged at base.
All paint work in poor order.
Crack in concrete floor.

2.4 Yard and access road

Hardcored road and yard. Very uneven surface with many depressions and several potholes, with a cluster of these at the entrance. Ditch alongside road overgrown and blocked. No protective fencing. Pair of 10' metal entrance gates, which do not shut properly and need re-hanging. Side rails on E side missing.
Fence against O.S. No. 561 belongs away.

We, the undersigned, having inspected the holding DECLARE the above to be a true and accurate record of the condition of the same as at the date taken.

.... B.Sc F.R.I.C.S. F.A.A.V.
Chartered Surveyor
Acting on behalf of
H. J. Scriven Esq., Landlord.

.... F.R.I.C.S. F.A.A.V.
Chartered Surveyor
Acting on behalf of
Flatmead Farming Partnership, Tenants.

Note: Photographic records should also be identified, initialled and signed, as being accurate.

CHAPTER TWENTY-THREE
Livestock

23.1 Valuation of livestock is a complex subject, and it is accepted by those in the livestock industry that some people have natural ability, whereas others have not. Nevertheless, constant handling of stock in markets and noting types, weight and prices, will teach most people the rudiments of livestock valuation, even if they do not possess natural flair. It is important to have a thorough knowledge of the class of stock being valued, as there is a world of difference between valuing, say, Friesian dairy cattle and Beef stores. Many of the larger firms of livestock auctioneers have specialists in their own fields to deal with different classes of stock. This chapter can only touch upon the rudiments of the subject.

2.3.2 BEEF CATTLE

Currently, the most popular beef cattle in demand are the continental crosses i.e. progeny of native breed cows bulled with continental breeds, perhaps in this order of merit: Limousin, Charolais, Simmental, Blonde D'Aquitaine, Gelbvich, Belgian Blue and Marchigiana. The reason is because these crosses produce well fleshed carcases, often of good conformation, and produce a good percent age of saleable lean meat. However, it should be noted that these continental type cattle spend a long period of their lives growing, and as a rule do not fatten off grass unless they are fed concentrate or are out of pure beef cows, such as Herefords, Angus or Welsh Blacks. Pure Friesian steers are still very popular, provided they are out of good Friesian stock and do not have much Holstein blood. Holstein dairy cattle have become very popular in recent years, but they are a large, rangy, lean type of cattle, which have generally poor conformation and for beef production there is much difficulty experienced by beef producers in getting these cattle finished.

Hereford crosses also are still popular, provided they are not crossed with Holstein cows. Pure bred Hereford steers are also in good demand and sell very well, but Heifers are not popular because they are too small and carcases tend to be very fat. Aberdeen Angus cattle are not popular with beef producers because they are usually small and do not return a good weight for age. The beef produced from these pure beef breeds—particularly steer beef—is in good demand but in many cases tends to be too fat, especially heifer beef. As the fat has to be trimmed off, there is much waste—consequently the butcher is not prepared to pay high prices for this class of beef. Shorthorn beef is not popular. The ideal beef animal has a body shaped like a brick—well-fleshed, firm, has good length with well developed hind quarters and loins, where the high priced cuts are found. It should not have a large belly.

It is a fairly easy task to a livestock auctioneer, farmer, butcher or dealer who is well versed in his job to value beef cattle. A quick assessment is made of the breed of animal, its conformation, its condition and its age. The valuer will assess the weight in kilogrammes (unless the animal has been weighed as in a market) and that is multiplied by the current price per kilo obtainable for that class of animal.

23.3 DAIRY CATTLE

It is accepted in the profession that valuing dairy cattle is a far more difficult task than valuing beef cattle. The modern Holstein Friesian has become, universally, the most popular dairy animal in the U.K. The Ayrshire and Dairy Shorthorns, once popular, have largely disappeared from whole counties. The Channel Island breeds—Jersey and Guernsey—are not popular with many because of the poor bull calves they produce, and also their limited cull value.

In valuing dairy cattle, the following matters need consideration:

(a) Breed.
(b) Whether pedigree or not. If so, it is registered with the Breed Society?
(c) If pedigree—what is the breeding? Some blood lines are extremely popular, particularly cows sired by well-known proven bulls. Most pedigree breeders are fully aware of the most popular blood lines.
(d) Age? Old cows with a limited productive life are not wanted unless they have a good pedigree or are, perhaps, in calf

to a well-known bull and there is a possibility of having a heifer calf born.

(e) If recorded or not? Good records of milk produced enhance the value of a cow. Yields of over 6,000 litres in a normal lactation (say 305 days) is considered good.

(f) The milk is of good quality. Milk is now paid on compositional quality, based on fat, protein and lactose. (Average quality milk has 3.94% fat, 3.23% protein, 4.56% lactose). Also, milk produced in late summer/early autumn is more valuable than that produced in the peak grass period of May/June. This means that a producer selling high quality milk at the right period will get more than the norm for his milk and this obviously will be reflected in the value of dairy cows that meet this criteria.

(g) The cow is good on inspection. A good cow should have dairy-like qualities (contrast a very blocky beef type cow which rarely gives much milk) and should be in the shape of three triangles or wedges—looking from the rear end, sideways and from above. She should have a well formed spacious udder—stretching forward underneath and up towards the tail at the end. Teats should be well spaced, and there should be no spare teats. The teats should be free from warts. It should have good milk veins curling forward from the udder. The cow should be well marked and true to breed type.

(h) The animal must be healthy. Many cows have lost quarters (due possibly to mastitis), have cut teats or perhaps, if old, large udders almost touching the ground. Any defect of this kind reduces the value appreciably.

(i) It must have good conformation so that it will realise a good price when sold as a cull cow. Ayrshire and Channel Island breeds have lost popularity because they do not realise high prices as cull cows and do not produce good beef cross calves by comparison with Friesians.

(j) What stage the cow is in its lactation and, if served, when it is due to calve. Cows lying off calving a long time (stale) are obviously not as valuable as a cow that is due to calve or is freshly calved.

Having taken the above factors into count, valuations can only be made by direct comparison with other cows or herds sold. Even when whole herds are sold, there is great variation in values—one might find a good young commercial cows, just calved, realising say 1,000 guineas, whilst a few minutes later an older stale cow, not of particular merit, may only realise say 350 guineas.

23.4 DEFINITIONS ETC.

A 'slip' or 'Cow heifer' is a heifer that has aborted a calf and is usually fattened for the butcher. It will have developed an udder.

A '*bullock*' or '*steer*' is male bovine, castrated. This is usually done with a bloodless castrator at the age of 2–3 months.

A '*three quarter*' cow or heifer produces milk from three quarters only—the other being blind.

Age of Cattle—Cattle have baby teeth until they are 21 months old. Two larger teeth then appear, followed by, for each year afterwards, two large teeth (permanent incisors) until they have eight large teeth at four years old.

'*Store Cattle*' are generally sold by their producers to others for growing further, and often finishing as beef animals.

'*Dual Purpose*' cattle are cattle kept both for milk and beef e.g. South Devon, Dairy Shorthorn and Red Poll.

23.5 SHEEP

This chapter does not deal with pedigree sheep, most flocks of which are kept for the purpose of producing ram lambs, or rams for sale, and also some ewes, or ewe lambs for sale to other breeders, who will buy them for breeding purposes.

The great majority of flocks kept are commercial flocks, kept for fat lamb production (in some cases production of store lambs) with production of wool a secondary matter.

The sheep industry has been revolutionised in the U.K. in the last decade or so. Many formerly popular breeds have lost their popularity because they no longer produce the type of fat lambs required by the meat trade, and in particular, meat exporters. The demand today is for a well-fleshed lean carcase, the most popular weight being from 16 to 19 kg estimated dead carcase weight. Very heavy fatty carcases are no longer wanted because the modern housewife does not want fat, and this has to be trimmed off at a considerable loss to the retailer. Many of the breeds producing heavy fat lambs have thus declined, in particular, the pure bred Down, Clun Forest, Kerry and other similar breeds. Nevertheless, they are still used by some flock masters as crossing breeds and, as such, produce a ewe that is in demand e.g. the English halfbred, which is the progeny of the cross of a Blue-faced Leicester ram on a Clun Forest ewe.

The Suffolk cross is still arguably the most popular lamb, since it is usually fat at the weight the trade demands and also has good conformation, a small head and fine legs. It does not get too fat and has little waste on the carcase.

Other very popular breeds are crosses, such as, Mule (Blue-faced Leicester ram × Swaledale ewe), Welsh halfbred (Border Leicester ram × Welsh ewe). Welsh Mule (Blue-faced Leicester ram x Welsh ewe or Brecon Cheviot ewe). Scotch halfbred (Border Leicester ram × Cheviot ewe). However Mules are not always popular as some have poor, narrow hindquarters.

There are also variations which are popular in certain areas, such as the Scotch halfbred (Border Leicester × Cheviot) which is in much demand in the Midland Counties—but the lambs tend to be of heavier weights at maturity.

Valuation of store lambs is a matter of experience, taking into account breed, condition, weight and the trend of the market at the time of valuation. Fat lambs can be fairly easily valued if the correct weight can be estimated and by reference to market prices and the guarantee payments at the time of valuation.

Valuing breeding ewes is more difficult. The following factors have to be considered:

(a) Breed.
(b) Type.
(c) Age. Ewes have only bottom teeth. Up to 12 months it has its suckling teeth. At one year it has two large teeth at two years—four, at three years—six, and at four years-eight. It is then called 'full mouth' and can exist as such for some time. Eventually, it loses teeth and then becomes 'broken mouth' meaning that it is aged and of much less

value, possibly of only cull value. However, an experienced person can tell if a ewe is old, despite having a so-called 'good mouth'.

(d) Udder. It is most important that a ewe should have a good udder, as otherwise it cannot rear its lambs well. The udder should be sound and when a ewe lambs, should be full of milk.

(e) Health. The condition of a ewe is important. To produce and rear a lamb well, it must be in very good health. Some flockmasters lose lambs (and ewes) at, or near, lambing time due to pregnancy toxemia (so-called twin lamb disease). This arises as a result of the pregnant ewe not being fed well enough to sustain herself and the lamb growing inside her. It is, therefore, important that pregnant ewes should have supplementary feed for at least a month prior to lambing (say $\frac{1}{4}$–1 kg ewe cobs a day). After lambing, many flockmasters creep feed their lambs for several weeks if early lambs are to be produced. In any event, ewes should be fed hay and concentrate, both before and after lambing, as grass is usually short then.

Valuations of ewes are made by direct comparison with the realisation prices of comparable ewes, taking the above into account.

Notes

(i) Ewes are usually injected about 2–3 weeks prior to lambing, against Dysentry, Pulpy Kidney, Blackleg, etc. This results in both ewe and lamb being protected.

(ii) Young male lambs are castrated (usually by the rubber ring method) at about 2–3 weeks. Some flockmasters also cut the end of their tails at the same time.

(iii) Unless sold fat off their mothers, lambs are generally weaned at about 12–14 weeks of age.

(iv) Ewes and lambs sold together are called 'couples'.

(v) Fat sheep are sold in £ & p per head. Prior to sale time, they are inspected by a Meat and Livestock Commission grader, in the market, and provided they satisfy his requirements, they are accepted for certification and he will give the 'estimated dressed carcase weight (E.D.C.W.) which is announced prior to the sheep being sold by auction.

23.6 PIGS

This section (by Alan N. Lane F.R.I.C.S., F.A.A.V.) will deal with the valuation of commercial pigs as opposed to pedigree pigs. Just as pig keeping has become more of a specialised, intensive system of farming over the last 20 years so has the valuation of this class of stock. Many markets who have traditionally sold pigs for many years do not do so now and therefore the opportunity for young valuers to see pigs sold by auction is declining.

As with beef and lamb the housewife's demand for lean meat and the farmer's demand for a quick growth rate has resulted in the rise in popularity of certain pure breeds and hybrids and the decline in popularity of some of the older breeds used in years gone by to produce big, and often fat carcases. Examples of pure breeds remaining popular are the Large White, the Welsh and the Landrace and those now declining to the ranks of rare breeds include the Gloucester Old Spot, Tamworth and Berkshire.

The butcher is looking for a long carcase to produce pork chops and a rounded hind quarters to produce legs of pork and hams, and this is achieved often by the crossing of popular breeds. Feeding is also an important factor in achieving the desired end product.

The valuer will often encounter three classes of pigs:

(a) *Breeding Stock*. This includes breeding gilts, sows and boars. With a breeding sow or gilt, conformation is important but also the general health and wellbeing. She should be well fleshed but not excessively fat and have good feet and legs. She should have an adequate number of well placed teats—14 is ideal. There should be no sign of mastitis or other infection.

A stock boar will usually be of pure breed and again must have good general health, sound feet and legs and not be fat and lazy.

Important factors affecting valuation of sows or gilts in farrow are the farrowing date, the number of litters she has produced and the boar by which she has been served. A sow close to farrowing will be worth more than one only recently served. The 'average' sow produces six litters in a lifetime, so when valuing a sow who has had, say, four or five litters her likely cull value is an important factor in establishing a value. With young gilt or sow her potential to produce another four or five litters must be regarded.

Often a breeding sow will be valued with her litter of pigs which may be up to six or eight weeks old. They are valued as a family, the overall value depending on the condition and size of the sow, the number, evenness of size and general health of the piglets.

(b) *Store Pigs.* This description covers the range from weaned piglets up to approximately 40 kg liveweight. Pigs are weaned generally at six to eight weeks old (although some intensive systems wean at 21 days old) and are commonly referred to as weaners up to the age of 10 to 12 weeks old when they may weight up to 30 kg liveweight. Factors affecting valuation include breed, size, conformation and general health and vigour. Small pigs intensively reared and spending most, if not all, of their life inside, are particularly susceptible to disease although modern veterinary techniques and environmentally controlled buildings reduce this to a minimum. Often baby piglet's tails are cut at birth. This is a plus factor when valuing store pigs as risk of tailbiting causing consequent stress and injury is reduced. Male piglets castrated when young sell better as stores as they can be mixed with gilts without causing stress. Some fatteners will buy uncastrated pigs, particularly if selling at pork or cutter weights before the pigs reach sexual maturity. Valuation is usually by direct comparison.

However, many feed compound firms operate weaner groups organising the direct movement of weaners from the breeder to the fattener. Often these pigs are bought unseen by the buyer who probably takes regular supplies from one or two farms and who relies on them being of a consistent standard. These pigs often change hands at a payment based on the gross weight of the load of pigs.

(c) *Fattening Pigs.* Pigs upwards of 40 kg liveweight being fed for slaughter are usually referred to collectively as fattening pigs. The sale of fat pigs is not covered by any form of guarantee scheme as steers or lambs, the last such scheme having ended in the early 1970s. However, the weight range descriptions originating from those days are still used today and the Meat and Livestock Commission still record prices achieved for fatstock market report purposes, based on these weight ranges: pork pigs, 40 kg–67 kg; cutter pigs, 68 kg–82 kg; bacon pigs, 83 kg–101 kg; overweights, 102 kg + .

Fattened pigs sold liveweight are weighed and sold in pence per kilogram liveweight, consequently the valuation of this class of stock is often by visual assessment of weight multiplied by a value in pence per kilogram. Factors affecting the value per kilogram include the weight (porkweight pigs are usually worth more per kilo than baconweight pigs), and most importantly the conformation.

CHAPTER TWENTY-FOUR

Arbitration

24.1 Many agricultural valuers will sooner or later experience cases that cannot be settled and in order to resolve the matter arbitration will be necessary. The law on Arbitration is fairly extensive and there are many legal cases, decided, which are relevant. However, this Chapter is intended to be a brief guide of the practical application of the law.

24.2 Arbitration was intended to be a relatively cheap method of solving such disputes but in recent years has proved to be in some cases a somewhat more expensive method of deciding issues than originally intended. It is therefore very important in clients interest that every effort should be made to settle valuations and disputes without resorting to arbitration.

24.3 AGRICULTURAL ARBITRATIONS OPERATE UNDER TWO CODES:

Tenancy Disputes—Agricultural Holdings Act 1986

Other matters—e.g. a dispute on the valuation clause on the sale of a farm where the Arbitration Acts 1950 and 1979 govern the issue.

24.4 TENANCY DISPUTES

Section 84 of the Agricultural Holdings Act 1986 provides for arbitration of matters required to be determined under the Act or the Regulations made under it. Schedule 11 of the 1986 Act sets down the arbitration procedure to be adopted in such proceedings.

24.4.1 Appointment Procedure

When an issue arises, e.g. rental revision, failure to agree an outgoing tenants claim, or a landlords counterclaim for dilapidations (in the latter two cases where 8 months have elapsed since the termination of the tenancy), it is usual to agree the name of a person to act as an arbitrator. It is normal for one party to submit to the other three names of possible arbitrators and if any are acceptable that person is chosen and appointed. However, in practice, it is sometimes even difficult to achieve agreement on any one name. In such cases an application, on the prescribed form (see example) enclosing the prescribed fee (at the time of writing £70) is made to the President of the Royal Institution of Chartered Surveyors. When this is received by the President notification of the application is given to the opposite party or his agent and if the case is one in Wales, the arbitrator must possess a knowledge of Welsh agricultural conditions and, if either party requires, a knowledge of the Welsh language.

24.4.2 The President will make an appointment from a panel appointed by the Lord Chancellor and from that date the person appointed takes on his role as arbitrator. In due course when this appointment is made the parties are notified by the President.

24.4.3 Arbitrators Procedure

On receiving his appointment the Arbitrator will write to the agents appointed by the Parties (or if no agents names are given to the parties direct) notifying them of his appointment, the date of his appointment and requesting the delivery to him, in duplicate, within 35 days of his appointment, statements of their respective cases with all necessary particulars.

It should be noted from paragraph 7(a) of Schedule 11 of the Act that no amendment or addition to the statement or particulars delivered will be allowed after the expiry of the said thirty five days except with the consent of the arbitrator. Furthermore, paragraph 7(b) of the same schedule provides that the party to the arbitration shall be confined at the hearing to the matters alleged in the statement and the particulars delivered by him.

24.4.4 The Statement of Case

Rental Arbitrations
Schedule 2 of the Agricultural Holdings Act 1986 sets out the matters which an Arbitrator must take into account in assessing the rent payable in respect of a holding.

The statement of case should therefore be drafted with the matters which the Arbitrator has to take into account in mind remembering the provisions of paragraphs 7(a) and (b) provisions of Schedule II already referred to. Prior to the passing of the Agricultural Holdings Act 1984 (now consolidated in the 1986 Act) the tendency amongst many valuers was to deliver rather brief and sketchy statements of case. This is an unwise course to adopt and it is advisable that a great deal of detail should be included in statements of case.

The statement of case should therefore set out the following:

1. Introduction

This should:

(a) give the name of the owner, the name of the tenant, the address of the holding, describe briefly the holding, the type it is and give its area.
(b) State the date the tenancy commenced, the rent then passing, subsequent revisions and also the current rent, and when this was fixed.
(c) If additional land was included in the tenancy give the date of such inclusion and also the additional rent payable.
(d) State when the notice requiring arbitration was served and the date (if applicable) when the application was made to the President of the R.I.C.S. for the appointment of arbitrator. It should also state the date on which the arbitrator, giving his name, was appointed.

2. Terms of the Tenancy

Detailed reference should be made to the tenancy agreement, the term date, dates when the rent is payable, if in advance or arrears, repairing liabilities, responsibility for insurance and any other matters in the tenancy agreement that are considered to be relevant, especially any unusual covenants.
A plan should be provided.

3. Character and Situation of the Holding

The situation of the holding should be given and also a description of the holding provided with details of the residence, cottages (how occupied) the buildings and the land. Reference should be made to the topography, the type of soil including references to soil types (obtained from the Soil Survey), quality (grading of soil) and its crop bearing capacity. Also availability of water supplies, altitude of the land, accessability, etc.

4. Productive Capacity

Details should be given of the farming system practised, including particulars of stock kept, crops grown and if so considered, give the opinion that the system practised is the best for the holding. If production cannot be increased this should be stated.

5. Relating Earning Capacity

Here reference to a budget for the next year should be made to illustrate the earning capacity that can be expected to follow the productive capacity. It may not be convenient to provide a budget at this stage and if so it should be stated that this will be provided at the hearing. This budget should be in great detail and should show the profits to be expected to be earned. Ideally the budget should be for the next three years, but in practice an arbitrator will be content with the first years budget, well knowing the volatility of supply and demand for agricultural products and resultant up and downs of prices and consequently the profits that can be expected to be earned.

NOTE: It is advisable, to save the arbitrators time at the hearing to have prior agreed the budget with the agents opposite number, but in practice it is rare for a budget to be totally agreed.

6. Level of Rents of Comparable Lettings

If good information, supportive of the case is available, reference to specific comparisons to be cited should be made, (in practice, it is difficult except in the case when the holding is on an estate when the landlord has the information, to provide comparisons. Other tenants rarely like their business to be bandied about and are reluctant to divulge the necessary information or to give permission for their farms to be inspected). If comparisons are to be given, the name and address of the comparable farm(s) should be provided including all the terms of the tenancy, rent passing and when last revised, tenants improvements etc., and the tenancy agreement should be provided. An opportunity should be given for the opposing party to inspect the comparison *prior* to the hearing and also to inspect the tenancy agreement.

If no comparison is to be given, reference here can be made to Ministry of Agriculture Rent Reports, or other evidence available cited, in a general way, in order to illustrate rental trends.

7. Tenants Improvements

Under this heading a detailed list of tenants improvements (executed with or without landlords permission) should be given, including dates carried out. It is unnecessary to give costs (which historically may be low). The rental value equivalent of the tenants improvements should be given (but it is not necessary to provide details of the calculation at this stage)

If a claim is being made by the Landlord for additional rent in respect of landlords improvements (discounting grant aid received) this should be stated, giving full details of the improvements, and, in this case the amount of additional rent sought.

8. Rental Valuation

The rental considered to be properly payable and sought to be awarded should be stated, but at this stage it is not necessary to provide details of how the valuation is calculated.

9. General

Mention should be made of the following, if applicable, and dependent in whether it is the Landlords or Tenants Statement of Case:

9.1 The demand for holdings and scarcity of holdings to let.
9.2 Any particular disadvantages, in a general way the holding suffers from and what abnormal difficulties or problems may be encountered in farming the holding.
9.3 Any special features and advantages the holding possesses.
9.4 Any lack of fixed equipment.
9.5 Any other matter of relevance to the issue of the rent which is properly payable.

10. Costs

A statement should be made that the arbitrator will be addressed on costs.

24.4.5 The Hearing

When both statements of case are received, the arbitrator will exchange them and the hearing will then take place. The venue for this is often a room at the farmhouse, a local inn or village hall. The arbitrator will arrange the seating, having his legal advisor (if any) at the top table, and the parties facing each other at a table in front of the arbitrator. A place, normally facing the arbitrator, will be reserved for witnesses to give evidence.

The arbitrator is really a judge of the matters at issue and decides the issues on the evidence before him. It is important that he should remember this and control the proceedings in a dignified, formal and courteous manner but at all times he should be in complete control. He should not allow smoking. Strangers who have nothing to do with the proceedings should not be allowed to attend as the proceedings are private. The arbitrator should act in a totally impartial manner and should not show favour to anyone. An offer of coffee or luncheon made (it frequently is so) should be refused as the receipt of any form of hospitality from one party may be strongly objected to by the other.

The hearing procedure more or less follows Court actions and is as follows:

1. *The Claimant* (or rather his advocate) opens his case and provides evidence.
2. He calls his witnesses who are put on oath by the arbitrator.
3. Witnesses are cross examined.
4. Witnesses may be re-examined.
5. *The Respondent* (or rather his advocate) opens his case and provides evidence.
6. The respondents witnesses are called to give evidence on oath.
7. Witnesses are cross examined.
8. Witnesses re-examined.
9. The respondent sums up his case.
10. The claimant sums up his case.
11. *Costs* The claimant or his advocate address the arbitrator on costs and this is followed by the respondent making an address on costs.

The advocate/valuer usually produces a detailed, written proof of evidence and budget for his client, and all witnesses are also advised to have written proofs of evidence. The witness after being sworn reads his proof of evidence. In this way important points and evidence are not missed out, and time is saved. Copies of the witnesses proof of evidence are supplied to the Arbitrator and the other party after the witness is sworn.

The arbitrator will be taking notes of the evidence given, by receiving proofs of evidence and documents during the proceedings. Sometimes the proceedings are recorded or a stenographer is present. If a valuer is acting as both an advocate and giving evidence, the arbitrator will request him to take the oath or affirmation.

24.4.6 The Inspection

Some arbitrators make it their practice to make a brief inspection, alone, before the hearing is held especially in rental arbitrations in order to acquaint themselves with the property concerned. This is much to commended as the arbitrator will have the advantage of knowing his subject much better than where he is a total stranger to the property.

Inspections are normally made at the conclusion of a hearing or if there is insufficient time, on another day. Most arbitrators prefer to make inspections unaccompanied unless it is essential that the parties or their valuers should be present to point out matters which were the subject of an issue at the hearing. If one party is present, then the other or his representative should also be there to avoid any chance of undue representations being made, in his absence, to the Arbitrator.

24.4.7 The Award

The award must be issued within 56 days of an Arbitrators appointment, unless the time has been extended, on application of the arbitrator with the consent of the parties to the President of the Royal Institution of Chartered Surveyors.

The Award must be in the form laid down by the Agricultural Holdings (Form of Award in Arbitration Proceedings) Order 1990 (No. 1472). The Award must state separately, the amounts awarded in respect of separate claims or matters of a claim e.g.:

tenant right matters
disturbance claim,
dilapidation claim

It is not necessary to show individual amounts for specific claims unless a request has been made by one of the parties.

The award must clearly allocate responsibility for the arbitrators costs and also the parties costs. Furthermore it must give a date, not later than one month from the date of delivery of the award for the payment of compensation and costs. Awards should be dated, signed and witnessed, and also state to whom they were delivered and on what date.

It is very important that an award should be clear in its intent, should deal finally with the matter (unless it is an interim award) and must be decisive and not open to question.

24.4.8 Reasons for the Award

Arbitrators are frequently requested to give reasons for their award and do so, if requested, under the provisions of the Tribunals and Inquiries Act 1971, Section 12. Such requests must be made on or before the giving or notification of the decision. Great care is needed in drafting reasons as what appears to be a wrong reason can be a ground for seeking the Courts to set aside an award or to remit the award or any part of it to the reconsideration of the Arbitrator.

Special Case

An arbitrator may at any stage of the proceedings and shall if so directed by the County Court, state a special case for the opinion of the County Court on an question of law arising in the arbitration or as to his jurisdiction.

24.4.9 Costs

The costs of, and incidental to, the arbitration and award shall be at the discretion of the arbitrator. Such costs can on the application of either party, be taxable in the County Court.

The arbitrator shall in awarding costs take into consideration the reasonableness or unreasonableness of the claim, of either party, and any unreasonable demand for particulars or refusal to supply such and generally the circumstances of the case.

It is now widely held that costs should follow the event, i.e. the loser should pay. This is in sharp contrast with the practice adopted extensively in the past by many arbitrators, of dividing costs. However a 'Calderbank Offer' (*Calderbank* v. *Calderbank* [1975] 3 All ER 333) may have a major bearing on the award of costs that have arisen after the offer is made. A Calderbank offer can be done by submitting to the opposite party a written offer in settlement which is made 'Without Prejudice except as to costs'. This offer is made *well in advance* of the hearing and if not accepted, the Arbitrator at the hearing is handed a sealed envelope which is marked 'to be opened after the award, but before determination of costs.' If the arbitrators award is the same or less than the Calderbank offer, then he will normally award to the party the costs which have arisen since the offer was made to the other party.

24.4.10 Procedural Points at the Hearing

1. The Arbitrator opens the proceedings by issuing an appearance list for completion by those attending.
2. He states who he is and that he is acting as an Arbitrator and then should state who has appointed him, whether the parties or the President of the R.I.C.S., the date of his appointment and also state the matter which is referred to his decision.
3. He then asks the advocate/valuers for their formal appointments to act.
4. He reminds the parties of the normal procedure that is adopted in such hearings.
5. When witnesses are called, he administers the oath or affirmation.
6. He should examine all documents submitted in evidence and see that the opposite party has a copy of the same. In particular where documents should be stamped, he should not receive them in evidence unless they *are* stamped or the cost of stamping together with the penalty for late stamping is deposited there and then to the arbitrator.
7. If an adjournment is sought this should be granted unless it is clearly a time wasting and stalling device. However, the view of the opposing party should always be sought before the arbitrator decides whether or not to grant an adjournment.

8. He should fix a time for the inspection.
9. He should obtain the counter signature of the parties to the arbitrators application (if one is to be made) for an extension of time in which to issue his award beyond 56 days from his appointment.

24.5 ARBITRATIONS OTHER THAN RENTAL

There are many other matters regarding the subject of agricultural arbitrations, other than rental, and these include:

Succession on retirement—the rental payable and the terms of the tenancy.

Succession on death—the rent payable and the terms of the tenancy.

Terms of a new *written tenancy agreement* where one did not previously exist.

Reduction of rent following the landlord obtaining possession of part of the holding.

Damage by Game—compensation payable.

Settlement of claims by outgoing tenants for tenant right, improvements and fixtures.

Settlement of the landlords claim for dilapidations and deterioration.

24.5.1 Statement of Case

Again, as in the case of rental arbitrations there is no official form of statement of case prescribed, but broadly the statement should, apart from giving details of the tenancy refer to the tenancy agreement, give the names of the landlord and tenant, the termination date of the tenancy (where relevant), show the amount of the claim in detail, (where applicable), give the section(s) of the Act under which the claim is made and where applicable, e.g., in the case of a dilapidation or tenant right claim, give fully itemised particulars of the claim showing each separate item of claim and the amount of claim against each item.

24.6 EXAMPLE RENTAL ARBITRATION

OLD HENDRE, ROSS ON WYE is a 403.01 acre old red sandstone arable farm with Georgian farmhouse, modern Woolaway 3 bed bungalow, one other cottage (let) and a good set of buildings including 500 ton grain store erected by the tenant 6 years ago. The soil is easy working, Grade II (60%), Grade III (40%) on the Land Classification Map. The farm has extensive frontage to the River Wye (which floods 80 acres) and the tenant has obtained an irrigation licence.

It is owned by the Aral Estate Co and is let to Mr. J. Yeomans on an annual Candelmas (2nd February) tenancy which commenced 10 years ago. Mr. Yeomans obtained the tenancy by open tender and at the time paid a rent of £18,135 per annum on a full repairing and insuring agreement. During the negotiations he contended that when he obtained the tenancy of the holding he added an overbid of 25% of the rent payable as key money, in order to secure the tenancy. The rent has remained unaltered since. The landlord served on 28th January 1989 a notice under Section 12 of the Agricultural Holdings Act 1986 requiring arbitration on the rental. The landlords agents have been seeking a revised rental of £22,165 per annum. The tenant has, reluctantly offered an increase of £1,000 per annum which has been refused. The parties have failed to agree an arbitrator and an application was made by the landlord, on 3rd January 1990 to the President of the R.I.C.S. for an arbitrator to be appointed. The President appointed Mr. I. A. Wright FRICS, FAAV as arbitrator on 28th January 1990.

The farming system practised is growing 100 acres winter wheat, 80 acres winter barley, 40 acres spring barley, 35 acres potatoes, 65 acres sugar beet and the remaining 80 acres is pasture keeping 400 breeding ewes. 100 beef bulls are produced annually. 500 root tegs are produced having been purchased in November of each year as stores. Mr. Yeomans is 46 and keeps 2 men.

TENANCY AGREEMENT

Dated 4th February 1980. A normal tenancy agreement except that the Tenant carries out all repairs and pays landlord fire insurance premium.

TENANTS IMPROVEMENTS

Erection of 500 tonne grain store. Tile drained 20 acres. Installed central heating in farmhouse. Installed 3 phase electricity supply.

AGRICULTURAL HOLDINGS ACT 1986

APPLICATION FOR THE APPOINTMENT OF AN ARBITRATOR FOR USE IN S.12 RENT CASES)

To the President, The Royal Institution of Chartered Surveyors, (Arbitrations Section), 12 Great George Street, Parliament Square, London. SW1P 3AD.

A. The landlord and the tenant having failed to agree as to the person to act as arbitrator and there being no provision in any agreement between them relating to the appointment of such arbitrator. We West and Stone, 6 High Road, Ross on Wye, Herefordshire, HR9 6EQ, hereby apply to the President of the Royal Institution of Chartered Surveyors for the appointment of an arbitrator (from among the Lord Chancellor's panel of arbitrators) as to the rent to be paid for the holding referred to below as from the next termination date following the date of the demand for arbitration served by the landlord on 28th January 1989 on his tenant.

B. We, West and Stone enclose a cheque for £70 made out to 'The Royal Institution of Chartered Surveyors'.**

C. We, West and Stone understand that unless the application, accompanied by the fee, is received at the address given above before the next termination date following the date of the demand for arbitration, it will be invalid and the appointment of the arbitrator will not be made.

Signature ......West and Stone...... Date ......3rd January 1990......

State whether landlord or tenant or duly authorised agent of either

Agent for the Landlord ......Agents for the Landlord......

PARTICULARS REQUIRED	REPLIES
1. Name and address of the agricultural holding (as defined in Section 1 of the Agricultural Holdings Act (1986).	Holding(s): Old Hendre Parish: Ross on Wye County: Hereford & Worcester
2. Name and address of landlord.	Aral Estate Co., c/o Agents West & Stone
3. Name and address of landlord's agent (please quote reference).	West & Stone, 6 High Rd, Ross on Wye, Herefordshire
4. Name and address of tenant.	Mr. J. Yeomans, Old Hendre, Ross on Wye, Herefordshire
5. Name and address of tenant's agent (please quote reference).	Mr. J. T. Thomas FRICS, FAAV, FSVA, The Hollies, Ross on Wye
6. Approximate area of holding.	403 Acres
7. Description of holding (for example, mixed, arable, dairying, market garden).	Mixed arable & stock farm
8. Has a demand in writing for an arbitration as to the rent to be paid for the holding been made by one party to the other? If so, state the date of such demand and whether it was made by the landlord or the tenant.	Yes on 28th January 1989 made by the Landlord
9. State the next termination date following the date of the demand.	2nd February 1990
10. (a) On what date did the tenancy commence.	2nd February 1980
(b) On what date did any previous increase or reduction of the rent take place?	No alteration in rent since 2nd February 1980
(c) On what date took effect any previous direction of an arbitrator that the rent should continue unchanged?	N/A

11. (Holdings in Wales) Do you require the appointment of an arbitrator with the knowledge of the Welsh language. N/A

NOTES

** VAT is not payable and the fee is non-returnable. If more than one holding is to be referred to arbitration, a fee is payable (and a separate form must be submitted) in respect of each holding.

Under paragraph 7 of the Eleventh Schedule to the Agricultural Holdings Act 1986, the parties to the arbitration must, within 35 days from the appointment of the arbitrator, deliver to him a statement of their respective cases, with all necessary particulars. Such a statement cannot afterwards be amended, except with the arbitrator's consent; and the parties will be bound by it when the hearing takes place.

I. A. Wright Esq FRICS FAAV.,
Chartered Surveyor,
2184 High Town,
Hereford.

To: Messrs West & Stone, 6 High Road, Ross on Wye.
Mr. J. Thomas, Chartered Surveyor, The Hollies, Ross on Wye.

Dear Sir,

AGRICULTURAL HOLDINGS ACT 1986 HOLDING:
OLD HENDRE FARM, ROSS ON WYE, HEREFORDSHIRE
LANDLORD: ARAL ESTATE CO.
TENANT: MR. J. YEOMANS

I write to confirm that I have been appointed by the President of the Royal Institution of Chartered Surveyors to act as Arbitrator in this case to determine the rent properly payable for the above holding as from the next termination date following the demand for arbitration made by the landlord to the Tenant. My appointment is dated 28th January 1990.

1. Statements of Case

Statements of Case must be submitted to me in duplicate by each party within 35 days of the date of my Appointment. On receipt of both, copies of each will be sent to the other party.

2. Date of Hearing and Place

It would be helpful if you could suggest a convenient date, about 6/7 weeks hence, for the hearing. Also a suggested venue for the hearing.

3. Representation

Please let me know if the parties will be legally represented at the Hearing either by Counsel or by a Solicitor. Please also let me know how many witnesses, if any, each party will be likely to call.

4. Costs

My costs in this matter will be £60 for opening the file and then at the rate of £75 per hour including time spent considering representations, travelling etc., plus all disbursements and out-of-pocket expenses and VAT. I would ask both parties to confirm agreement to this before the Hearing.

5. Generally

May I remind you that all communications with me should be in writing, and that both parties should ensure that any letters written to me should be copied and simultaneously despatched to the other party.

I would also remind the parties that no inadmissible evidence should be produced and no 'Without Prejudice' discussions, communications or negotiations should be referred to in any way if they have not resulted in any agreement.

I look forward to hearing from you on this matter as soon as possible.

Yours faithfully,

I. A. Wright

AGRICULTURAL HOLDINGS ACT 1986

IN THE MATTER OF ARBITRATION BETWEEN:

ARAL ESTATE CO. (Landlord)

and

MR. J. YEOMANS (Tenant)

Concerning the rent to be properly payable in respect of the Holding known as:

OLD HENDRE, ROSS ON WYE

In the County of HEREFORD & WORCESTER

STATEMENT OF CASE for

THE ARBITRATOR with particulars

on behalf of the Landlord.

---

1. INTRODUCTION

Old Hendre Farm, situated in the Wye Valley, 4 miles south of the market town of Ross on Wye is a 403 arable and stock farm owned by the Aral Estate Co. and is tenanted by Mr. J. Yeomans.

2. TENANCY

The tenancy commenced on 2nd February 1980 and the rent payable at that time was £18,135 per annum. There has been no subsequent rental revision in the intervening 10 years and the rental passing is still this sum. A notice requiring arbitration was served under recorded delivery by the Landlords Agents on 28th January 1989.

Subsequently, negotiations have taken place firstly between the landlords agents and the tenant direct, and as matters could not be agreed an application was made on 3rd January 1990 to the President of the Royal Institution of Chartered Surveyors to appoint an arbitrator and Mr. I. A. Wright, FRICS FAAV was appointed on 28th January 1990 by the President to determine the issue.

3. TERMS OF THE TENANCY

The holding is held under a written tenancy agreement dated 4th February 1980 (a copy of which is annexed). This is a normal tenancy agreement (which will be provided in evidence) except that the tenant has undertaken to be responsible for all repairs and the payment of the landlords fire insurance premium.

4. CHARACTER AND SITUATION OF THE HOLDING

Old Hendre is a very good quality arable and mixed farm well situated on a council road 4 miles south of Ross. It is capable of growing most kinds of cash crops. It has a good quality 5 bedroom Georgian residence, extensive modern and traditional buildings and good quality land with soil being derived from the old red sandstone formation being Grade II and III soil on the Land Classification map. The land is either level or gently undulating and a substantial area has frontage to the River Wye.

The holding has two cottages, one occupied by the tenants workman and the other sub-let by him (with landlords permission).

5. PRODUCTIVE CAPACITY AND RELATED EARNING CAPACITY

The holding is primarily used for arable production but a ewe flock and a bull beef enterprise are run from the farm. Wheat, barley, potatoes and sugar beet are grown. The holding has a 1105 tonnes 'A' beet quota and a 14.5 ha potato quota.

A budget illustrating the productive and thus related earning capacity of the holding will be prepared and submitted in evidence to the arbitrator.

6. LEVELS OF RENTS AND COMPARABLE LETTINGS

Evidence will be given of a comparable letting on the Aral Estate and the general level of rents now payable not only on this Estate but also regionally and nationally. Reference will be made to Ministry of Agriculture published rental statistics.

7. TENANTS IMPROVEMENTS

The tenant has carried out the following improvements:
7.1 Erection of a 500 tonne grain store.
7.2 Drained O.S. 560, 20.12 Acres.
7.3 Installed 3 phase electricity.
7.4 Installed central heating in the farmhouse. It is however, observed that whilst 7.1–7.3 were beneficial to the holding 7.4 was for the tenants personal comfort.

8. GENERAL

Reference will be made to:
8.1 The general scarcity of holdings to let and the extensive demand for tenancies.
8.2 The value of the Beet and Potato quota the farm possesses.
8.3 The trends and other factors affecting farm rents.
8.4 All other matters considered relevant to the rent of the holding.

9. RENTAL VALUATION

Taking into account all the above matters and having regard to current market conditions and the terms of the tenancy, we consider that the rental properly payable for the holding under the provisions of S.12 and Schedule 2 of the Agricultural Holdings Act 1986 and taking into account all relevant factors, is in the sum of £22,165 (Twenty two thousand one hundred and sixty five pounds) per annum.

10. PLAN

A plan of the holding will be provided at the hearing.

11. EVIDENCE

The right is reserved to enlarge on the above at the hearing and witnesses will be called as appropriate to give evidence.

12. COSTS

The right is reserved to address the arbitrator on costs.
Dated this 28th day of February 1990.

Signed ....

J. R. Andrews FRICS FAAV
for West and Stone,
Agents for the Landlord,
Chartered Surveyors and Land Agents
6 High Road, Ross on Wye,
Herefordshire. HR9 6EQ

E & O E

AGRICULTURAL HOLDINGS ACT 1986

IN THE MATTER OF ARBITRATION BETWEEN

ARAL ESTATE CO. (Landlord)

and

MR. J. YEOMANS (Tenant)

Concerning the rent of the property payable in respect
of the Holding known as:

OLD HENDRE, ROSS ON WYE

in the county of HEREFORD & WORCESTER

STATEMENT OF CASE for
THE ARBITRATOR with particulars
on behalf of the Tenant.

---

1. INTRODUCTION

Old Hendre Farm, is a 403 mixed arable and stock farm situated 4 miles from Ross on Wye, 1 mile off the A40 road and is in the Wye Valley. It is tenanted by Mr. J. Yeomans from the landlords the Aral Estate Co.

2. TENANCY

The annual candlemas agricultural tenancy commenced on 2nd February 1980 at a rental of £18,135 per annum. A notice requiring arbitration on the rental was served by the Landlords Agents on 28th January 1989.

Subsequent negotiations have failed to resolve the matter and the President of the Royal Institution of Chartered Surveyors appointed Mr. I. A. Wright FRICS FAAV as arbitrator to decide the issue of 28th January 1990.

3. TERMS OF TENANCY

The holding is held under the terms of a written tenancy agreement dated 4th February 1980. The tenancy commenced on 2nd February 1980 and is an annual tenancy with rental paid half yearly, in arrears on 2nd August and 2nd February. It is a standard tenancy agreement excepting that the tenant has the abnormal burden of being responsible for all repairs and paying the costs of fire insurance of the fixed equipment.

4. CHARACTER AND SITUATION OF THE HOLDING

The holding is primarily an arable farm which is either level or part undulating and a large part of the land is subject to flooding by the River Wye. It has a *substantial* residence and adequate buildings, part modern, part traditional. There are also included a pair of cottages.

5. PRODUCTIVE CAPACITY AND RELATED EARNING CAPACITY

The arable enterprise comprises: growing wheat, barley, potatoes, sugar beet, and *approx.* 80 acres of grassland (liable to flooding) maintains the 400 breeding ewes kept. About 500 root tegs are fattened in winter and there is also a bull beef unit with an output of about 100 head annually.

A budget illustrating the productive capacity and thus the related earning capacity of the holding will be produced at the hearing. Also the past three years farm accounts will be provided.

6. GENERAL LEVEL OF RENTS

The latest Ministry of Agriculture rental statistics will be produced and referred to, showing the virtual standstill of rentals of holdings of this nature.

7. TENANTS IMPROVEMENTS

The tenant has carried out the following improvements at his sole expense:

7.1 Erected steel and concrete 500 tonne grain store.
7.2 Installed 3 phase electricity supply.
7.3 Tile drained 20 acres—O.S. No. 560.
7.4 Installed oil fired central heating in the farmhouse—19 radiators.

8. GENERAL

Reference will be made at the hearing to the following relevant matters affecting the rental payable for the holding:

8.1 The constantly diminishing level of profits experienced by farmers and the tenant, in particular, in recent years.
8.2 The withdrawal and general reduction of guarantee payments for both the arable and livestock sector.
8.3 Constantly increasing fixed and variable costs incurred.
8.4 The increasing pressure mounted on agriculture be the conservation and green element of the community and general 'farmer bashing' which ultimately affects demand and increases costs. The extensive use by the public of the 'Wye Valley' footpath along this farm results in losses and nuisance from trespass.
8.5 When the farm was originally let, the rental paid, by tender, was at least 25% above the level of rents, paid for similar holdings. A substantial 'Key' element was included in the tender.
8.6 Any other relevant matters.

9. RENTAL VALUATION

Having taken into account all the above and other relevant matters it is considered that the rent properly payable for the holding, having regard to the onerous terms of the tenancy is £19,135 per annum.

10. EVIDENCE

The right is reserved to fully amplify the above statement at the hearing and witnesses will be called in evidence.

11. COSTS

The Arbitrator will be addressed on costs.

Dated this 25th day of February 1990.

Signed ....

J. T. Thomas FRICS FSVA FAAV
Chartered Surveyor,
Agent for the Tenant,
The Hollies,
Ross on Wye,
Herefordshire.

AGRICULTURAL HOLDINGS ACT 1986

FORM OF AWARD

Arbitrator: IAN ARTHUR WRIGHT F.R.I.C.S. F.A.A.V., 2184 High Town, Hereford
Date of Appointment: 28th January 1990
Time for making award extended to: 30th May 1990
Present Landlord: Aral Estate Co, c/o West & Stone, 6 High Road, Ross-on-Wye
Present Tenant: Mr J. Yeomans, Old Hendre, Ross-on-Wye
Rent payable prior to arbitration: £18,135 per annum

AWARD OF THE ARBITRATOR

The claims or questions set out in the Schedule to this award have been referred to arbitration and, having considered the evidence and the submissions of the parties, I, the arbitrator, award as follows.

1. As from 2nd February 1990 (the next day on which the tenancy could have been brought to an end by notice to quit at the date of the notice demanding arbitration under section 12 of the Act) the rent previously payable is increased to £20,140 per annum being the rent properly payable in respect of the holding at the date of the reference to arbitration.
2. The cost of and incidental to the arbitration and the award shall be dealt with as follows:

(a) My costs of the award amounting to £1,660 plus VAT of £249 must be paid by the landlord and the tenant in the following proportions:

The Landlord £996 plus VAT of £149.40
The Tenant £664 plus VAT of £99.60

(b) As respects the costs of and incidental to this arbitration the Landlord must bear his own costs and must pay 20% of the costs of the tenant to be taxed in the county Court as prescribed by the County Court Rules.

(c) All costs ordered by me to be paid shall be paid on the 28th day after delivery of this award.

Signed by the arbitrator I. A. Wright in the presence of: J. Anderson, 2184 High Town, Hereford.

Date: 30th April 1990

This award was delivered to the Landlords Agents: Messrs West & Stone on the 6th day of May 1990.

THE SCHEDULE

	Sum Awarded
Rent	£20,140 per annum

FORM OF AWARD prescribed by The Agricultural Holdings (Form of Award in Arbitration Proceedings) Order 1990 (SI 1472)

AGRICULTURAL HOLDINGS ACT 1986

FORM OF AWARD

Arbitrator: [name and address]

Date of appointment:

Time for making award extended to:

Present landlord: [name and address]

Present tenant: [name and address]

Rent payable prior to arbitration:

AWARD OF THE ARBITRATOR

The claims or questions set out in the Schedule to this award have been referred to arbitration and, having considered the evidence and the submissions of the parties, I, the arbitrator, award as follows:

1. The landlord is to pay to the tenant in respect of the claims set out in Column 1 of Part I of the Schedule the sum(s) set out in Column 2 thereof.

2. The tenant is to pay to the landlord in respect of the claims set out in Column 1 of Part II of the Schedule the sum(s) set out in Column 2 thereof.

3. As from (the next day on which the tenancy could have been brought to an end by notice to quit given at the date of the notice demanding arbitration under section 12 of the Act) the rent previously payable [is [increased] [reduced] to £] [continues unchanged at £] being the rent properly payable in respect of the holding at the date of the reference to arbitration.

4. The notice to quit referred to in Part IV of the Schedule shall [not] have effect. [I postpone the termination of the tenancy until].

5. My award in respect of the claims set out in Column 1 of Part V of the Schedule is set out in Column 2 thereof.

6. The landlord must pay to the tenant the sum(s) awarded by me to the tenant on the day after delivery of this award, and the tenant must pay to the landlord the sum(s) awarded by me to the landlord on the same day.

7. The costs of and incidental to the arbitration and the award shall be dealt with as follows:

(a) My costs of the award amounting to £ must be paid by the [landlord] [tenant] [landlord and the tenant in the following proportions]:

(b) As respects the costs of and incidental to this arbitration [each party must bear his own costs] [the landlord must pay [% of] the costs of the tenant] [the tenant must pay [% of] the costs of the landlord] [to be taxed in the County Court] [according to Scale [ ] as prescribed by the County Court Rules] [[the landlord [the tenant] must pay £ to the [tenant] [landlord] on account of his costs]:

(c) All costs ordered by me to be paid shall be paid on the day after delivery of this award.

Signed by the arbitrator in the presence of:

Date:

This award was delivered to the [landlord] [tenant] on [date].

THE SCHEDULE

Part I

Column 1 *Claims made by the landlord*	*Column 2* *Sum(s) awarded*

Part II

Column 1 *Claims made by the tenant*	*Sum awarded*

Part III

RENT

Part IV

Question(s) arising out of a notice to quit

Part V

Column 1 *Other claims*	*Column 2* *Award*

24.7 ARBITRATIONS OTHER THAN UNDER THE AGRICULTURAL HOLDINGS ACT

Other disputes arise which are not tenancy matters, e.g. disputes on a valuation of produce, crops etc, on the sale of a farm. These disputes are not resolved under the agricultural arbitration code, but under the provisions of the Arbitration Acts of 1950 and 1979. Certain arbitrations relating to milk quotas are governed by the Agriculture Act 1986.

Where an issue cannot be resolved the contract will usually provide that the matter shall be referred to an arbitrator and it is usual for the contract to state who appoints the arbitrator.

The procedure, which will be in the total control of the arbitrator is not dissimilar to a normal agricultural tenancy dispute procedure excepting that there are no particular time limits laid down, except as stipulated by the arbitrator.

A usual procedure to be adopted is as follows:

1. The arbitrator may call 'Preliminary Proceedings' if he considers this necessary, and will inform the parties of the procedures to be adopted, date of submission of claims, replies, etc. If these proceedings are not called he will, otherwise write to the parties and inform them of the procedure. He has absolute discretion.
2. A 'Points of Claim' will be requested from the Claimant within a given time, say one month.
3. The respondent will then be given say a further 14 to 28 days to submit a 'Points of Defence' statement.
4. The arbitrator may be asked to order the delivery of 'Further and Better Particulars of Points of Claim'.
5. A further period may be given for a 'Reply'.
6. The hearing will then be held, much as in the case of an agricultural tenancy arbitration and in due course, the Award will be available on payment of the Arbitrators Costs.

CHAPTER TWENTY-FIVE

Milk Quotas: Compensation, Procedure and Rent Reviews

(By P. D. CARTER M.A. (Oxon) F.R.I.C.S., F.A.A.V.)

25.1 This chapter is concerned only with that milk quota legislation which relates specifically to milk quota registered in the name of a tenant of an agricultural holding.

25.2 COMPENSATION

After the introduction of the milk quota system into the United Kingdom on 2nd April 1984 it soon became apparent that quota had acquired a value and that tenants of agricultural holdings with quota should, on termination of their tenancies, be able to share in the value of that quota. Section 13 of the Agriculture Act 1986 gave the right to tenants to receive compensation for milk quota in accordance with the provisions set out in Schedule 1 to the Act.

It should be noted that a tenant does not own quota although it is registered in his name. The tenant's interest in the capital value of quota is confined to the amount of compensation that he will be entitled to receive at termination.

It should be noted that 'holding' in the Agriculture Act 1986 has the same meaning as in the Dairy Produce Quotas Regulations, that is 'all the production units operated by the producer and located within the geographical territory of the Community'. In this Chapter the expression 'agricultural holding' means that area of the producers holding which is subject to a tenancy in respect of which:

(a) a claim is to be assessed under the Agriculture Act 1986 Section 13 (compensation to outgoing tenants for milk quota)
(b) there is a reference under Section 12 of the Agricultural Holdings Act and the rental value of transferred quota is to be considered under the Agriculture Act 1986 Section 15 (rent arbitrations: milk quota)

25.3 The aim of the provisions in Schedule 1 is to entitle tenants whose tenancies are terminating to receive compensation for the value of the following quota:

(a) *Transferred Quota (TQ)*
Quota that has been acquired from another producer and transferred to the agricultural holding insofar as the tenant has borne the cost of the transaction which effected the transfer.

(b) *Excess Quota*
All quota which, after deduction of transferred quota, exceeds the standard quota (SQ) for the agricultural holding. This reflects the extent to which the tenant's management resulted in more quota being allocated to the agricultural holding than standard quota.

(c) *Tenant's Fraction (TF)*
A proportion of standard quota to reflect the contribution made by the tenant towards the provision of dairy improvements and fixed equipment.

25.4 THE STEPS IN OUTLINE

In order to prepare a claim it will assist a valuer to break the claim down into a number of steps as follows:

(1) Does the tenancy qualify?
(2) Does the tenant qualify?
(3) Calculate the number of litres of quota registered in the tenant's name which are relevant to the agricultural holding by apportionment if necessary to give relevant quota (RQ).
(4) Calculate how much of the relevant quota is relevant transferred quota (TQ).
(5) The balance of (RQ–TQ) will be relevant allocated quota (AQ).
(6) Assess the relevant number of hectares (RH).
(7) Assess the reasonable amount (RA).
(8) Select the appropriate prescribed quota per hectare (PQH) and prescribed average yield per hectare (PAYH) from the current Milk Quota (Calculation of Standard Quota) Order and calculate standard quota (SQ).
(9) Adjust SQ for any shortfall in quota allocated to the tenant following an award by the Dairy Produce Quota Tribunal.

(10) Assess the annual rental value (ARV) of qualifying tenant's dairy improvements and fixtures.
(11) Assess the proportion of rent paid (PRP) during the relevant period in respect of areas used for feeding, accommodating and milking dairy cows.
(12) Calculate tenant's fraction (TF).
(13) Assess the proportion of TQ (from 4) which qualifies for compensation.
(14) Calculate the number of litres for compensation (QFC).
(15) Assess the value of quota and total compensation due.

25.5 THE STEPS IN DETAIL

The steps in detail are as follows:

(1) **Does the tenancy qualify?**
Qualifying tenancies are:

(a) a tenancy under which the tenant enjoys security of tenure by virtue of Section 2 of the Agricultural Holdings Act 1986.
(b) a tenancy for a term of years approved by the Minister
See Agriculture Act 1986 Schedule 1 paragraph 18 (1) definition of 'tenancy'

Excluded tenancies are:
(a) a tenancy for a period of between one and two years, a *Gladstone* v *Bower* tenancy.
(b) a grazing or mowing tenancy for a period of less than one year.

(2) Does the tenant qualify?

(a) the tenant must have quota registered as his. Many tenants farm their holdings through the medium of a partnership or a company and the registered producer is not the tenant himself. Such arrangements may disqualify a tenant from claiming compensation and valuers should advise that a tenant's name appears on the register before a claim is made.
(b) the quota must be registered in relation to land of which the agricultural holding either is the whole or a part. Whilst in practice this is unlikely to cause difficulties, tenants whose

registered address is remote from the agricultural holding should be aware of the requirement.

(c) the tenant must have had quota allocated to him, in relation to the agricultural holding, under the Dairy Produce Quotas Regulations 1984; and if the tenant is claiming compensation for quota transferred at his expense and registered in his name, he may claim compensation in respect of that quota provided that he either:

(d) had quota allocated to him as in (c),
or:

(e) was in occupation of the agricultural holding as a tenant (though not necessarily under the present tenancy) on 2nd April 1984.

See Agriculture Act 1986 Schedule 1 paragraph 1 (1).

Other tenants who are entitled to claim compensation are:-
Tenants who have succeeded to tenancies by virtue of the grant of a tenancy after 2nd April 1984:

(f) following a direction by the Agricultural Land Tribunal under Section 39 Agricultural Holdings Act 1986 (succession on death) or Section 53 (succession on retirement) or under corresponding previous legislation.

(g) following a direction under Section 39 Agricultural Holdings Act 1986 where a new tenancy is granted by agreement between landlord, tenant and potential statutory successor to that successor.

(h) to a tenant who was a 'close relative' of the previous tenant by agreement with the landlord under Section 31 (1) (b) or (2) Agricultural Holdings Act 1986.

See Agriculture Act 1986 Schedule 1 paragraph 2.

Tenants who are assignees, where a qualifying tenancy has been assigned after 2nd April 1984 (whether by deed or by operation of law). Any allocated quota or transferred quota which would fall to be assessed for compensation in the hands of the assignor should be treated as having been allocated or transferred to the assignee.

See Agriculture Act 1986 Schedule 1 paragraph 3.

In the case of sub-tenancies, a head tenant, who sublets to a sub-tenant who is entitled to claim compensation from the head tenant, can claim a like amount from the head landlord.

See Agriculture Act 1986 Schedule 1 paragraph 4.

Tenants who may not claim compensation will include those who took occupation of the agricultural holding after 2nd April 1984 and did not have quota allocated to them. Thus tenants who have transferred quota or have paid a premium to a landlord for quota and have taken occupation after 2nd April 1984 are barred from statutory compensation.

(3) Establish the relevant amount of registered quota which is to be assessed (RQ)

(a) where the agricultural holding is the only land farmed by the tenant, then registered quota will be the relevant quota.

(b) where the agricultural holding forms only part of the land farmed by the tenant, the registered quota must be apportioned as at the termination date between the agricultural holding and the balance of the land farmed.

Since all those persons who have an interest in the land farmed by the tenant have a right to make representations to an arbitrator appointed to determine an apportionment, both landlord and outgoing tenant should ensure that all interested parties agree that the apportionment is in accordance with areas used for milk production as specified in the transfer form. Failure to do so may trigger off arbitration.

In practice, a tenant will be well advised to follow the procedures leading to prospective apportionment not more than six months preceding the termination of the tenancy. This should avoid an arbitrator appointed to determine compensation being delayed by having to determine apportionment which may have to be determined at another arbitration, at the worst, by another arbitrator.

Although a strictly mathematical apportionment takes no account of land quality or investment, Article 5.2 of Commission Regulation (EEC) No. 1371/84 suggests that the mathematical approach is correct in that it requires quota to be apportioned between the agricultural holding and other land farmed 'in proportion to areas used for milk production'.

The expression 'areas used for milk production' was brought to the attention of the High Court in *Puncknowle Farms Limited* v *Kane* [1985] 3 All ER 790. The Court ruled that the expression included the following:

> 'the areas of the set of buildings and yards belonging thereto used for the production of milk and forage areas used by the dairy herd and to support the dairy herd by the growing of grass and any fodder crop for the milking dairy herd, dry cows and all dairy following female young stock (and home bred dairy or dual purpose bulls for use on the premises, if applicable) if bred to enter the production herd and not for sale. In this case, maize, silage, hay and grass were the fodder crops, but consideration would have been given to corn crops, or part of corn crops, grown for consumption by the dairy herd or young stock, including the use of straw, had agreed evidence been produced.'

This very wide definition applies only for the purposes of apportionment under the Dairy Produce Quota Regulations and must not be confused with other similar expressions used in the Agricultural Act 1986.

> *See Agriculture Act 1986 Schedule 1 paragraphs 1 (1) and 1 (2) for definition of 'the relevant quota'. Dairy Produce Quotas Regulations 1989 as amended Regulation 10 (Apportionment of Quota) and Schedule 4 (Apportionment and prospective apportionments by arbitration).*

(4) Calculate how much of relevant quota is relevant transferred quota (TQ)

Transferred quota and allocated quota are apportioned to the agricultural holding on the same basis. A tenant cannot, therefore, seek to retain all transferred quota on that part of the land he farms which is not the agricultural holding in respect of which he is claiming compensation.

(5) Assess relevant allocated quota (AQ)

Deduction of the relevant transferred quota from relevant quota will give the relevant allocated quota (AQ).

(6) Assess the relevant number of hectares (RH)

The definition of 'the relevant number of hectares' in the Act is:

(a) 'the average number of hectares of the land in question used during the relevant period for the feeding of dairy cows kept on the land, or'

(b) 'if different, the average number of hectares of the land which could reasonably be expected to have been so used (having regard to the number of grazing animals other than dairy cows kept on the land during that period)'

The relevant period is the period in relation to which allocated quota was determined. In most cases wholesale quota was allocated on the basis of the number of litres delivered to the Milk Marketing Board (or other purchaser) during the period 1st January 1983–31st December 1983.

Since it is mandatory to use the second definition (b) if it is different to the first definition (a), it is logical to ignore (a) and proceed with (b). If (b) is the same as (a) so be it!

The expression 'land in question' refers to the agricultural holding in respect of which compensation is being claimed.

The reference to 'land used for the feeding of dairy cows kept on the land' is further defined as excluding 'land used for growing cereal crops for feeding to dairy cows in the form of loose grain'. This curious definition is unclear but possibly is meant to exclude land used for growing cereals which were used under definition (a) or could reasonably be expected to have been used under definition (b) for home mixed concentrates. The land used for feeding dairy cows will clearly include grazing land, land used for growing fodder crops for conservation and areas where cows were fed, such as silage clamps and parlour standings with feeders. Other areas which it may be argued fall to be considered are roadways, tracks and other areas which facilitate feeding.

Dairy cows are defined as 'cows kept for milk production (other than uncalved heifers)'. A qualifying dairy cow starts as a newly calved heifer and finishes once she has completed her final lactation. Thus a barren cow kept by a tenant after the has ceased milk production to put on weight for slaughter cannot qualify as a dairy cow but will fall to be considered under definition (b) as a 'grazing animal other than a dairy cow'.

The reference to 'kept on the land' is unclear. There are instances were no dairy cows were in fact ever physically present on the holding during the relevant period but fodder was taken from the holding to feed dairy cows kept on other land farmed by the tenant. There are probably more cases where the holding is bare land serviced by fixed equipment elsewhere. In such circumstances valuers should bear in mind definition (b) and consider the average number of hectares which could reasonably be expected to have been used during the relevant period for the feeding of dairy cows kept on the land. An objective assessment of the fixed equipment and the land comprised in the agricultural holding may suggest that dairy cows might reasonably have been expected to have been kept on the land where there was the fixed equipment to support them. On the other hand where there was no fixed equipment, then it may be argued that it would not have been reasonable to expect a tenant to have used the agricultural holding for feeding dairy cows.

Definition (b) requires a valuer to have 'regard to the number of other grazing animals . . . kept' on the holding. The interpretation of this subjective test can be argued in at lease two ways.

1. Assess the average number of hectares (a) which could reasonably have been expected to have been used for feeding dairy cows. Discover the actual number of other grazing animals kept on the agricultural holding during the relevant period and assess the average number of hectares (b) that might reasonably have been expected to have been used during the relevant period for feeding that number. Area (a)–area (b) will be the relevant number of hectares (RH).
   or:
2. Discover the actual stocking rates during the relevant period and use these as a guide to assessing the potential stocking rates when assessing the area which could reasonably have been expected to have been used for feeding dairy cows during the relevant period.

The valuer is required to assess the average number of hectares. This requires a calculation of the areas of the holding used for other purposes than feeding dairy cows and the periods for which they were so used.

Areas which are subject to mixed stocking can be calculated on a livestock unit basis. Areas which are shut up should be considered on the basis of their next use. An area of grass shut up for cutting for silage for feeding cows would thus be included.

The use of keeper sheep in the winter months can be argued on one hand to be included on the basis that the purpose of the sheep is to improve grassland for the spring growth for feeding dairy cows, and on the other hand it may be argued that the use of the area is for feeding sheep and not dairy cows.

It is convenient, to summarise the definition of the relevant number of hectares (RH) as:

> 'the average number of hectares of the holding which could reasonably be expected to have been used during the period in relation to which allocated quota was determined for the feeding of dairy cows kept on the holding (an objective test), having regard to the number of other grazing animals kept on the holding during that period (a subjective test)'.

See Agriculture Act 1986 Schedule 1 paragraph 6 (1) (a) 'the relevant number of hectares' paragraph 8 'the relevant period'.

(7) Assess the reasonable amount (RA)

The reasonable amount is the number of litres of milk which, by virtue of the quality of the agricultural holding and the climatic conditions in the area, could reasonably be expected to have been produced from one hectare of the agricultural holding during the relevant period.

The valuer must consider all those factors, including atypical weather conditions such as drought or flooding, which will enable him to assess the level of milk production which could reasonably have been expected from the agricultural holding during the relevant period. It should be noted that the assessment does not require the valuer to have regard to other grazing animals.

(8) Select the appropriate prescribed quota per hectare (PQH) and prescribed average yield per hectare (PAYH) and calculate standard quota (SQ)

(a) Prescribed quota per hectare (PQH) is that number of litres which the Minister from time to time (to take account of national adjustments to registered quota) prescribes as being appropriate for three different classes of dairy cow and three different land types.

(b) Prescribed average yield per hectare (PAYH) is that number of litres which the Minister has prescribed as being appropriate for the same three different classes of breeds of dairy cow and three different land types as for prescribed quota per hectare. The average yields are those applying to the relevant period and are set out in the current Milk Quota (Calculation of Standard Quota) Order.

(c) Valuers should note the arbitrary dates on which amendments to the Milk Quota (Calculation of Standard Quota) Orders come into effect. They do not coincide with either the quota year or most traditional term dates.

(d) The formula for assessing standard quota is:

$$SQ = RH \times \frac{RA}{PAYH} \times PQH$$

Where:

RH	is the relevant number of hectares.
PAYH	is the prescribed average yield per hectare
RA	is the reasonable amount
PQH	is the prescribed quota per hectare

See Agriculture Act 1986 Schedule 1 paragraph 6 (2) 'the reasonable amount'.

(9) Adjust standard quota for a shortfall of quota allocated following an award by the Dairy Produce Quota Tribunal

Where the tenant made an application to the Dairy Produce Quota Tribunal and the Tribunal made an award of quota to the tenant, it is likely that not all the award was allocated to the tenant. In those cases standard quota is reduced by the full amount of the shortfall.

Where relevant quota is less than registered quota, then the shortfall must be similarly proportionally reduced.

See Agriculture Act 1986 Schedule 1 paragraph 6 (3).

(10) Assess the annual rental value (ARV) of qualifying tenant's dairy improvements and fixtures

ARV is the annual rental value at the end of the relevant period of the tenant's dairy improvements and fixed equipment on land used for the feeding, accommodating or milking of dairy cows kept on the agricultural holding. The annual rental value is that amount which would fall to be disregarded under paragraph 2 (1) of Schedule 2 of the Agricultural Holdings Act 1986 on a reference to arbitration as to rent properly payable as at the end of the relevant period.

A valuer will therefore sieve each item which he has to consider as follows:

(a) Did it exist at the end of the relevant period?
(b) Was it on land actually used at some time during the relevant period for feeding, accommodating or milking of dairy cows?
(c) Was it a dairy item?
(d) Was it fixed equipment as defined in Section 97 Agricultural Holdings Act 1986? It is immaterial for these purposes whether it is also a tenant's improvement.
(e) Did it contribute to an increase in rental value of the holding?
(f) Was there an obligation to provide it imposed on the tenant by the terms of the contract of his tenancy?
(g) What is the rent properly payable for the agricultural holding assessed in accordance with paragraph 1 of Schedule 2 to the Agricultural Holdings Act 1986 *rebus sic stantibus* (that is including *all* tenant's improvements and fixed equipment)?
(h) By how much should the rent assessed under (g) be reduced by reason of the disregard of increase in rental value of the item? That amount will be the ARV.

See Agriculture Act 1986 Schedule 1 paragraph 7 'tenant's fraction'.

Agricultural Holdings Act 1986 Schedule 2 paragraph 2 (1) Section 97 definition of fixed equipment.

(11) Assess the proportion of rent paid (PRP) in respect of the relevant period attributable to those parts of the agricultural holding used for the feeding, accommodating or milking of dairy cows

Whereas ARV is assessed on an arbitrated basis at the end of the relevant period, PRP is carved out of the rent actually paid for the agricultural holding in respect of the relevant period. Thus a benevolent landlord who was letting at a low rent will be at a disadvantage to a tenant who had just completed a dairy improvement scheme two days before the end of the relevant period.

(12) Calculate the tenant's fraction (TF) of standard quota

$$\text{TF is } \frac{\text{ARV}}{\text{ARV} + \text{PRP}}$$

(13) Assess the proportion of transferred quota which qualifies for compensation

If the tenant has borne all the cost of the transaction 100% of the transferred quota will qualify for compensation, if, say, only 50% then only 50% of the transferred quota will qualify for compensation.

See Agriculture Act 1986 Schedule 1 paragraphs 1 and 5 (3).

(14) Calculate the total number of litres of quota for which the tenant is entitled to compensation (QFC)

(a) Where allocated quota (AQ) is greater than standard quota (SQ):

$$\text{QFC} = \text{TQ} + [\text{AQ} - \text{SQ}] + [\text{TF} \times \text{SQ}]$$

(b) Where allocated quota (AQ) is less than standard quota (SQ):

$$\text{QFC} = \text{TQ} + \left[\text{TF} \times \text{AQ} \times \frac{\text{AQ}}{\text{SQ}} \right]$$

(15) Assess the value of quota and total compensation due

The value of a litre of quota which ranks for compensation is 'the value of the milk quota at the time of the termination of the tenancy in question and in determining that value at that time there shall be taken into account such [**all**] evidence as is available, including [**but not exclusively**] evidence as to the sums being paid [**at that time**] for interests in land.

(a) in cases where milk quota is registered in relation to the land; and
(b) in cases where no milk quota is so registered'.

There is usually a great deal of hearsay evidence available as to the value of quota derived from reports of quota sales. However since an essential element of such transfers will include either a sale of freehold land or a letting, valuers must carefully analyse the full value of land sales or lettings and the considerations paid for transfers of the quota. At arbitration, unless both parties agree evidence, a valuer must be prepared to submit full documentary evidence of transactions including commissions paid.

See Agriculture Act 1986 Schedule 1 paragraph 9.

PROCEDURAL AND OTHER MATTERS

25.6 Establishing Standard Quota and Tenant's Fraction during a tenancy

Where a tenant may be entitled to compensation for quota at the termination of his tenancy either the landlord or tenant may at any time before the termination of the tenancy by notice in writing served on the other demand that the determination of the standard quota for the land or the tenant's fraction shall be referred to arbitration.

Both tenants and landlords may wish to know standard quota and tenant's fraction during a tenancy. From the tenant's point of view it will enable him to have a clear indication of the number of litres which will qualify for compensation at termination. The procedure has the distinct advantage of ensuring that at termination he will receive compensation sooner than if standard quota and tenant's fraction have to be referred to arbitration after termination. From the landlord's point of view he will have a clear idea of how many litres he is likely to have to pay and in the event of his re-letting the farm, so that he can decide how he will deal with the incoming tenant. In that both standard quota and tenant's fraction require consideration of historical facts during the relevant period, the procedure may assist both tenant and landlord by dealing with the questions before facts are lost in the mists of time. In any event both tenant and landlord will be assisted by agreeing historic factual information even if that agreement is not crystallised by an agreement as to standard quota and tenant's fraction.

There is provision whereby any changes in circumstances between the time of any agreement in writing as to standard quota or tenant's fraction (or a determination by arbitration) and the termination of the tenancy can be taken into account by an arbitrator appointed to determine compensation at termination, provided that the circumstances are materially different.

Such circumstances will include a reduction in the area of the agricultural holding.

See Agriculture Act 1986 Schedule 1 paragraph 11 (6) and (7).

25.7 Notice of Intention to Claim

Written notice of the tenant's intention to make a claim under Section 13 Agriculture Act 1986 for compensation for milk quota must be served on the landlord within two months of the termination of the tenancy. Failure to serve such a notice is fatal and any subsequent claim is unenforceable.

See Agriculture Act 1986 Schedule 1 paragraph 11 (1).

There are however special rules applying where an application has been made for a direction following the death of a tenant and to circumstances where a tenant remains lawfully in occupation of the holding after termination of his tenancy. Such special rules are outside the scope of this chapter.

See Agriculture Act 1986 Schedule 1 paragraphs 11 (3) and (4).

25.8 Time limit for settling by agreement

Landlord and tenant have eight months from the termination of the tenancy to settle the claim in writing. After eight months, the claim must be referred to arbitration.

See Agriculture Act 1986 Schedule 1 paragraph 11 (2).

25.9 Arbitration procedures

Arbitration must follow the code set out in Schedule 11 of the Agricultural Holding Act 1986 except that the maximum period by which compensation and costs are to be paid after the delivery of the award is extended from one month to three months. There is no provision for the payment of interest.

See Agriculture Act 1986 Schedule 1 paragraph 11 (5).

Valuers acting for tenants in protracted disputes at arbitration should always bear in mind that an arbitrator may make an interim award, if he thinks fit.

See Agricultural Holdings Act 1986 Schedule 11 paragraph 15.

25.10 AN EXAMPLE

Richard Nightingale is tenant of two agricultural holdings:

(a) Hillview Farm extending to 60.5 hectares owned by William Buzzard. The farm has a modern dairy unit with accommodation for 120 Friesian cows and 30 young stock units.
and:

(b) Blackacre, some 10 hectares of bare land down to permanent pasture owned by Loamshire County Council used exclusively by Mr. Nightingale for grazing heifers.

Mr. Nightingale has 720,000 litres of quota registered in his name of which:

(a) 550,000 litres are primary allocated quota.
(b) 70,000 litres are secondary allocated quota following a development award of 100,000 litres.
(c) 100,000 litres are transferred quota acquired at the sole expense of Mr. Nightingale.

The relevant period was 1st January 1983 to 31st December 1983.

Mr. Nightingale has decided to retire and gave both his landlords notice to quit to take effect at Michaelmas 1990.

Calculate the amount of compensation for milk quota he is entitled to receive from each landlord using the steps 1–15 set out in 25.4.

(1) The tenancies

Both tenancies were granted over twenty years ago and therefore Mr. Nightingale had milk quota allocated to him in respect of both agricultural holdings.

(2) The tenant

Mr. Nightingale is a sole trader who has the milk quota registered as his.

(3) Calculate the relevant quota for each holding by apportionment as follows:

Hillview Farm : The areas used for milk production over the five years prior to Michaelmas 1990 have averaged 60 hectares each year.

Blackacre : The entire area of 10 hectares has been grazed for the five years prior to Michaelmas 1990 by young dairy stock. The quota is therefore apportioned 60/70 ths. to Hillview Farm and 10/70 ths. to Blackacre.

i.e. RQ for Hillview Farm $\frac{60}{70} \times 720{,}000 = 617{,}143$ litres

RQ for Blackacre $\frac{10}{70} \times 720{,}000 = 102{,}857$ litres

720,000 litres

The compensation in respect of each holding will now be assessed in turn.

Hillview Farm

(4) Calculate relevant transferred quota (TQ)

Transferred quota is 100,000 litres which is apportioned 60/70 to Hillview Farm.

$$TQ = \frac{60}{70} \times 100{,}000$$

$$= 85{,}714 \text{ litres}$$

(5) Calculate allocated quota (AQ)

$$AQ = RQ - TQ$$
$$AQ = 617{,}143 - 85{,}714 = 531{,}429 \text{ litres}$$

(6) Assess the relevant number of hectares (RH)

The base year is 1983. During 1983 Mr. Nightingale used some 50 hectares for feeding dairy cows and 10 hectares for feeding an average of 30 heifer yearlings and 30 heifer calves.

An assessment of Hillview Farm suggests that the fixed equipment of the farm is capable of supporting a dairy herd of 120 cows and that it would be reasonable to have expected that an average number of hectares required to feed 120 dairy cows during 1983 would have been 48.5 hectares, ie. RH = 48.5. This would leave some 11.5 hectares available for the young stock which were kept on the holding during 1983.

(7) Assess the reasonable amount (RA)

The fixed equipment of the farm will support 120 cows and an appreciation of the position, climate, land classification and the area suggest that a reasonable yield to expect from a Friesian cow would have been, in 1983, 5,500 litres. Thus the total amount of milk it would have been reasonable to expect to have been produced from Hillview Farm is 660,000 litres. Divide this amount by the total farm area of 60.5 hectares to give the reasonable amount which could reasonably have been expected to have been produced from one hectare of the farm.

$$RA = \frac{660{,}000}{60.5} = 10{,}909 \text{ litres per hectare of Hillview Farm}$$

(8) Select the appropriate PQH and PAYH and calculate SQ

The appropriate PQH and PAYH for Friesian cows kept on 'other land' are found in The Milk Quota (Calculation of Standard Quota) (Amendment) Order 1990 S.I. 1990 No. 48.

$$PQH = 7{,}670$$

$$PAYH = 9{,}000$$

$$SQ = RH \times \frac{RA}{PAYH} \times PQH$$

$$= 48.5 \times \frac{10{,}909}{9{,}000} \times 7{,}670$$

$$= 450{,}899 \text{ litres}$$

(9) Adjust SQ for any shortfall of quota awarded by The Dairy Produce Quota Tribunal

Mr. Nightingale received a development award of 100,000 litres of which 70,000 litres have been allocated, there is, therefore, a shortfall of 30,000 litres, of which 60/70ths or 25,714 litres is relevant to Hillview Farm

SQ is therefore adjusted 450,899 – 25,714 litres
= 425,185 litres.

(10) Assess the annual rental value of the tenant's dairy fixtures and improvements (ARV)

Mr. Nightingale had completed a Farm and Horticultural Development Scheme in 1983 in order to increase his herd from 100 cows to 120 cows. His dairy improvements and fixtures on land used for feeding and accommodating dairy cows were at 31st December 1983:

(i) Cubicles for 120 cows.
(ii) Slurry tower for three months slurry storage.
(iii) All parlour fittings, feeders and milking equipment.
(iv) A ten tonne bulk feed bin built into the parlour roof.

The rent properly payable for the farm is assessed *rebus sic stantibus* as at 31st December 1983 to have been £7,150.00 per annum. It is further assessed that a tenant would have been prepared to have paid £6,150.00 for the farm without the items (i) to (iv).

The ARV of the items is therefore:

£7,150.00 – £6,150.00 = £1,000.00 per annum.

(11) Assess the proportion of rent paid in respect of land used for feeding, accommodating or milking dairy cows (PRP)

During 1983 Mr. Nightingale paid a total rent of £5,250.00 for Hillview Farm. The land not used for feeding, accommodating or milking dairy cows included the farm house and garden, the farm cottage, and land and buildings used for feeding and accommodating his young stock. It is assessed that the rent should be apportioned £4,000.00 to the land used by dairy cows and £1,250.00 to the land used for other purposes.

Therefore PRP = £4,000.00

(12) Calculate the tenant's fraction (TF)

$$TF = \frac{ARV}{ARV + PRP}$$

$$= \frac{1,000}{1,000 + 4,000}$$

$$= .20$$

(13) Assess the percentage of transferred quota which qualifies for compensation

Mr. Nightingale acquired the transferred quota at his sole expense, therefore all the transferred quota qualifies for compensation.

(14) Calculate total number of litres of quota for compensation (QFC)

In this case AQ (531,429 litres) is greater than SQ (425,185 litres).

$$QFC = TQ + [AQ - SQ] + [TF \times SQ]$$
$$= 85,714 + [531,429 - 425,185] + [.2 \times 425,185]$$
$$= 85,714 + 106,244 + 85,037$$
$$= 276,995 \text{ litres}$$

(15) Assess the value of quota and total compensation due

Evidence suggests that the value of quota as at Michaelmas 1990 is 32.5p/litre.

Therefore total compensation due to Mr. Nightingale from William Buzzard is:

276,995 × .325 = £90,023.38

Blackacre

(4) Calculate relevant transferred quota (TQ)

$$\frac{10}{70} \times 100{,}000 = 14{,}286 \text{ litres}$$

(5) Calculate allocated quota (AQ)

$$AQ = 102{,}857 - 14{,}286 = 88{,}571 \text{ litres}$$

(6) Assess the relevant number of hectares (RH)

The land is bare land and under the circumstances it would not be reasonable to expect any dairy cows to be kept on the land. The assessment of RH is therefore nil. Further consideration of all those items leading to a calculation of standard quota is therefore unnecessary, as is a consideration of tenant's fraction.

(14) Calculate the total number of litres qualifying for compensation (QFC)

All the quota apportioned to Blackacre will therefore qualify for compensation since it is either transferred quota or allocated quota in excess of standard quota which is nil.

$$\begin{aligned} QFC &= TQ + [AQ - SQ] + [TF \times SQ] \\ &= 14{,}286 + [88{,}571 - 0] + [TF \times 0] \\ &= 102{,}857 \end{aligned}$$

(15) Assess total compensation due

Total compensation due to Mr. Nightingale from the Loamshire County Council is:

$$102{,}857 \times .325 = \underline{\underline{£33{,}428.52}}$$

25.11 RENT AND MILK QUOTA

Section 15 of the Agriculture Act 1986 directs an arbitrator, subject to any agreement between the landlord and tenant, to disregard any increase in rental value of the agricultural holding due to any relevant transferred quota which would fall to be apportioned to the agricultural holding on a change of occupation.

At an arbitration under Section 12 of the Agricultural Holdings

Act 1986, once an apportionment of registered quota has been agreed, the only quota subject to the above is transferred quota. Thus quota in excess of standard quota and the tenant's fraction of standard quota are not subject to the directions contained in Section 15 of the Agriculture Act 1986.

Assuming that the tenant's actual milk quota would be available to that hypothetical prudent willing and competant tenant the most satisfactory way to deal with milk quota at rent reviews is to consider the productive capacity (which is a matter of physical output expressed in tonnes or litres) and the related earning capacity in the light of that productive capacity. Related earning capacity will be affected by the number of litres of quota which a tenant has available in relation to the agricultural holding, in that he cannot exploit milk production in excess of that quota without incurring liability to levy.

See Agriculture Act 1986 Section 15 and Agricultural Holdings Act 1986 Schedule 2 paragraphs 1 (1) and (2).

25.12 AN EXAMPLE

It is convenient to consider an example based on a rent review as at Michaelmas 1990 for Hillview Farm.

The farm is capable of supporting and milking 120 cows. If a reasonable Friesian cow has a yield of 5,500 litres then the prudent, willing and competent tenant would expect to produce 660,000 litres of milk from Hillview Farm which is, insofar as milk production is concerned, the productive capacity of the farm.

It has, however, already been established that the quota relevant to the holding is 617,143 litres, of which 85,714 litres is transferred quota. Thus, for rental assessment purposes, the 'available' quota on the holding is only 531,429 litres. The related earning capacity of the holding must then be assessed in the light of this limitation.

The hypothetical prudent, willing and competent tenant, however, knowing that each cow had a potential yield of 5,500 litres, would also consider the following four options:

1. Lease in an additional 128,571 litres of quota to enable milk production to be increased to 660,000 litres without incurring levy.
2. Transfer in at his own expense an additional 128,571 litres.

3. Limit production to the available quota by reducing the number of cows to 96, each producing 5,500 litres. Utilise surplus forage areas (12 hectares) and accommodation for another enterprise, e.g. heifer rearing.
4. Limit production to available quota by restricting the production from his 120 cows to 4,428 litres each by feeding low levels of concentrates.

The effect which each option would have on the related earning capacity of the holding compared with the simple solution set out above, can be considered by the use of partial budgets to assess the extra annual profit each option would generate as follows:

Option 1

Extra Costs	£	*Extra Revenue*	£
Lease 128,571 litres quota @ 5.8p/litre	7,457	Sell extra 128,571 litres milk@ 18p/litre	23,143
Extra concentrates @ 0.05 kg/litre and £148/tonne	4,884		
Extra forage costs for higher yields of silage from 60 hectares @ £30/hectrate	1,800		
	14,141		23,143
Revenue Foregone		*Costs Saved*	
None	—	None	—
	14,141		23,143
Extra Profit	9,002		—
	£23,143		£23,143

Note that there is a risk to this option since the availability of quota to lease cannot be guaranteed for future years.

Option 2

Extra Costs	£	*Extra Revenue*	£
Pay back loan of £45,000 (35p/litre) over 5 years	9,000	Sell extra 128,571 litres milk @ 18p/litre	23,143
Interest @ 17% on average outstanding loan (i.e. £22,500)	3,825		
Extra concentrates (as above)	4,884		
Extra forage costs (as above)	1,800		
	19,509		23,143

Revenue Foregone		*Costs Saved*	
None	—	None	—
	19,509		23,143
Extra Profit	3,634		—
	£23,143		£23,143

Note that when after 5 years the loan is paid off, the extra annual profit increases to £16,459.

Option 3

Extra Costs	£	*Extra Revenue*	£
Extra concentrates @ 0.05 kg/litre and £148/tonne	3,933	Gross margin on 20 dairy replacement units @ £340	4,800
Extra forage costs for higher yields of silage from 48 hectares @ £30/hectare	1,440		
	5,373		4,800
Revenue Foregone		*Costs Saved*	
None	—	None	—
	£5,373		4,800
Loss	—		573
	£5,373		£5,373

Option 4

Extra Costs	£	*Extra Revenue*	£
None	—	None	—
Revenue Forgone		*Cost Saved*	
85,714 litres @ 18p/litre	15,429	Concentrates	3,256
Extra profit	—	Loss	12,173
	£15,429		£15,429

Having assessed the likely effect on profit which each option would have, and the risks associated with each, it is likely in this instance that the prudent, willing and competent tenant would transfer quota in at his own expense in order to secure his future profitability. The full cost of the purchase must be written off against profit since the quota will not be transferable away from the holding without the landlord's consent, albeit that it will rank for compensation in full at the end of the tenancy.

It could, of course, be argued that even if the hypothetical prudent, willing and competent tenant were to follow option 2, any increase in rental value deriving from quota transferred in at his own expense should, under the terms of Section 15 of the Agriculture Act 1986, be disregarded. This leads the valuer back to a consideration of option 1 where he must use his skill and experience to take into account the risk factor in arriving at the related earning capacity of the holding.

CHAPTER TWENTY-SIX

Soil, Climate and Productive Capacity

(by Paul Wright B.A. M.Sc.)

26.1. The quality and productive capacity of an agricultural holding is, to a large extent, governed by its soils and climate. Accurate soil assessment and the ability to interpret soil and climatic data for valuation purposes require training and practice that have often been neglected by Chartered Surveyors. The aims of this Chapter are to identify which soil and climatic data are significant, to describe what published information is available, and to present a system for assessing land quality and productive capacity. Finally, some case histories are presented in which the provision of precise soil and climatic data has played a key role in agricultural valuations.

26.2. Soil Maps and Agricultural Land Classification

In the late 1960's and early 1970's the Ministry of Agriculture, Fisheries and Food (MAFF) published a set of Agricultural Land Classification (ALC) maps of England and Wales at a scale of 1:63,360.These maps grade the land according to the severity of environmental constraints on agricultural production, taking into account such factors as soil, gradient, rainfall and altitude.There are five grades, the best being Grade I which is land with only very minor limitations, typified by Lincolnshire silt-land. Grade 5, with very severe limitations includes, for example, rough grazing at high altitudes in Wales and the West Country.

ALC maps were primarily intended as a planning tool, to identify and protect the best agricultural land, and this purpose they have served well. However, they are described by MAFF as 'provisional' because the amount of fieldwork that went into locating grade boundaries varied considerably from place to place. In some areas the grading was based on detailed ground observations of soil and topography, while elsewhere existing cropping patterns were the main guide. Thus, MAFF state that parcels of land of less than 80 hectares cannot be reliably graded using ALC maps alone, and further investigations are necessary. Moreover, the brief descriptive booklets accompanying the maps are insufficiently detailed to give the reasoning behind the grading of a particular piece of land.

Unfortunately, few users of ALC maps (apart from MAFF officials and planners) are aware of these limitations. Estate and land agents, in particular, have come to use the maps as a definitive statement of land quality in valuations for sale and rental purposes, but in many instances, such use is inappropriate and misleading. The valuer would arrive at a more accurate assessment by obtaining detailed soil, climatic and other site information specific to the land in question. This can then be translated into an accurate assessment of grading if required, using the revised guidelines published by MAFF (Agricultural Land Classification of England and Wales, 1988). Indeed, the ALC grading can be a useful summary of land quality provided it is properly researched and the reasons for a particular grading stated.

The 1988 revised version of the Agricultural Land Classification contains mainly the guidelines and criteria for land classification; the original ALC maps themselves have not been revised. The classification now has five Grades plus two Subgrades of Grade 3. The most productive and flexible land falls into Grades 1 and 2 and Subgrade 3a and collectively comprises about one-third of the agricultural land in England and Wales. About half the land is of moderate quality in Subgrade 3b or poor quality in Grade 4. The remainder is very poor agricultural land in Grade 5, which mostly occurs in the uplands.

Grade 1—excellent quality agricultural land

Land with no or very minor limitations to agricultural use. A very wide range of agricultural and horticultural crops can be grown and commonly includes top fruit, soft fruit, salad crops and winter harvested vegetables. Yields are high and less variable than on land of lower quality.

Grade 2—very good quality agricultural land

Land with minor limitations which affect crop yield, cultivations or harvesting. A wide range of agricultural and horticultural crops can usually be grown but on some land in the grade there may be reduced flexibility due to difficulties with the production of the more demanding crops such as winter harvested vegetables and arable root crops. The level of yield is generally high but may be lower or more variable than Grade 1.

Grade 3—good to moderate quality agricultural land

Land with moderate limitations which affect the choice of crops, timing and type of cultivation, harvesting or the level of yield. Where more demanding crops are grown yields are generally lower or more variable than on land in Grades 1 and 2.

Subgrade 3a—good quality agricultural land Land capable of consistently producing moderate to high yields of narrow range of arable crops, especially cereals, or moderate yields of a wide range of crops including cereals, grass, oilseed rape, potatoes, sugar beet and the less demanding horticultural crops.

Subgrade 3b—moderate quality agricultural land Land capable of producing moderate yields of a narrow range of crops, principally cereals and grass or lower yields of a wider range of crops or high yields of grass which can be grazed or harvested over most of the year.

Grade 4—poor quality agricultural land

Land with severe limitations which significantly restrict the range of crops and/or level of yields. It is mainly suited to grass with occasional arable crops, the yields of which are variable. The grade includes very droughty arable land.

Grade 5—very poor quality agricultural land

Land with very severe limitations which restrict use to permanent pasture or rough grazing, except for occasional pioneer forage crops.

For planning appeals in relation to development of land for building or mineral extraction it is usually necessary to employ a consultant who understands the ALC system to carry out a soil survey and reassess the grading of the land. MAFF will normally make their own re-survey where they are seeking to protect high quality agricultural land.

26.3. Soil Surveys

The organisation with responsibility for mapping the soils of England and Wales is the Soil Survey and Land Research Centre (formerly the Soil Survey of England and Wales), based at Silsoe, Bedfordshire. The Soil Survey of Scotland operates from the Macaulay Land Use Research Institute, Aberdeen. A soil survey has recently been established by the Department of Agriculture for Northern Ireland in Ulster.

The whole of mainland Britain is covered by published soil maps at a scale of 1:250,000 (1 inch = 4 miles) with accompanying Regional Bulletins describing the soils and land use. About 10% of England and Wales and most of lowland Scotland and some highland areas are mapped at a scale of 1:63,360 (1 inch = 1 mile) or 1:50,000. A further 15% of England and Wales is covered by detailed maps at 1:25,000 (2.5 inches = 1 mile). Addresses from which these maps can be obtained are given at the end of this Chapter.

26.4. Collecting Soil Information

Soil maps are usually made by soil surveyors who are graduates in the field of natural science such as geology or geography. With the aid of an auger and spade the surveyor examines the layers of soil to bedrock or rooting depth (100–120 cm). By making a series of such observations across the landscape they are able to produce a map showing the boundaries between different soils. Factors that are significant in making a soil survey are described in detail in the Soil Survey Field Handbook and include the following:

(i) *Geology*. This concerns the kind of parent material in which soils have developed and which usually controls their characteristics. Information about the geology is obtainable from maps produced by the British Geology Survey, whose address is given at the end of this Chapter. Geological maps come in 'Solid' and 'Drift' Editions. Solid geological maps show the location of rocks formed in pre-Glacial times such as Chalk, granite and London Clay. In many parts of Britain there is little or no relationship between the soils and the underlying rocks because a succession of glacial and interglacial episodes in the past two million years has blanketed the landscape with 'drift' deposits such as Chalky Boulder Clay, gravel and Brickearth. In many places, the soils have developed in these drift deposits and so it is the Drift Edition that is the most useful of the two, particularly where no soil maps are available. However, geological maps are not a satisfactory substitute for a soil survey. For example, in some districts the Lower Greensand outcrop on the geological map correctly depicts an area of light soils but these may range from barren sands to some of the best quality light loams.

(ii) *Colour.* The colour of a soil is a reflection of its natural drainage state, and can be used to make an assessment of the risk of waterlogging, even in summer, when the soil is dry. A permeable, well drained soil is a uniform brown or reddish colour throughout. Waterlogging produces grey and orange colours (termed 'gleying') that may range from a few mottles in the case of slight waterlogging to a predominance of grey or blueish-grey colours in severely waterlogged soils. Careful observation is required in soils in reddish parent materials as the grey coloration produced by waterlogging is less pronounced.

(iii) *Permeability.* Soil waterlogging has two principal causes. In low-lying situations with permeable soils such as fenland and river alluvium a groundwater-table rises through the porous subsoil in winter, usually falling again in summer. If adequate outfall or pumping facilities are available this situation can be permanently remedied by arterial drainage, and the agricultural land is then often of the highest quality. In Britain, however, waterlogging of soils is more commonly caused by the presence of impermeable clay or heavy loam in the subsoil that prevents percolation of rainwater, producing a 'perched' or 'surface' water-table in the upper layer of the soil. Improvement of such soils requires a pipe drainage system with permeable backfill and continuing secondary treatment in the form of moling and subsoiling or some combination of both. This type of waterlogging can never be fully relieved, and care will always be needed in autumn and spring to avoid compaction by stock and machinery.

(iv) *Texture.* This refers to the proportions of sand, silt and clay particles that make up the soil. Texture is usually assessed by moulding moist soil between finger and thumb. Accurate assessment requires training and constant practice.

A useful guide to texturing is the MAFF Leaflet No.895 Soil Texture.

(v) *Stones.* Excessive stoniness reduces the amount of water available to a growing crop, increases implement wear and interferes with drilling and harvesting. In extreme cases, stones and boulders may render land unimprovable.

(vi) *Depth.* Depth to rock, clay, gravel, sand or peat affects root development, drought risk and soil drainage.

(vii) *Organic matter* is a very important soil attribute. Peat, being moisture retentive and easily cultivated, produces some of the best horticultural soils in lowland areas. However, with drainage and cultivation, peat wastes away (at a rate of 1–2 cm a year on the Cambridgeshire fens) and so it is a diminishing asset on some farms. In high rainfall areas, a peat top soil is a disadvantage, being often acid and infertile, and because it has a low load-bearing capacity the land poaches easily, resulting in a shorter grazing season compared with adjacent non-peaty soils.

In mineral soils, particularly on light land, continuous arable rotations lead to a reduction in the organic matter content of the topsoil to an undesirably low level, reducing the water holding capacity and weakening the soil structure so that the surface slakes, caps and forms pans readily. Soil erosion by water may occur under these circumstances.

(viii) *Calcium carbonate content.* The natural calcium carbonate or free lime content of a soil affects its structure and pH level. The content of free lime in a soil is tested in the field by applying hydrochloric acid (10 percent solution) to a sample of soil. The calcium carbonate content of the Chalky Boulder Clays of East Anglia, such as in the Rodings of Essex, makes them better drained, more manageable and generally higher yielding than soils on acid London, Wealden or Lias Clays.

(ix) *Soil acidity.* Strong acidity (pH less than 4.5) is common in unimproved soils of upland areas. Even lower pHs of between 2.0 and 4.0 may occur in agricultural soils in lowland area when iron sulphide (pyrites) is present. This sulphide is most common in marine alluvium or peat and so extreme acidity is particularly associated with land in eastern England that is at or near sea level, but recently investigations have shown that it is not confined to these areas. The problem is caused by the oxidation of sulphide following drainage, and the production of sulphuric acid that causes the sharp fall in pH. At this level of acidity soil material is generally toxic to roots as soluble aluminium becomes available to plants.

Root development is impaired and crops wilt through drought stress on apparently water-retentive soils. Affected soil layers are usually in the subsoil, below the depth of routine lime testing and the problem can be sufficiently severe to reduce the sale or rental value of affected land.

(x) *Gradient.* The angle of slope affects use of machinery and there are certain critical gradients in the Agricultural Land Classification. For example, a slope of 1 in 5 is the upper limit for satisfactory cultivations and steeper land is placed in Grades 4 and 5.

All the above soil characteristics combine to give recognisable layers or *horizons* constituting the soil profile. It is the properties and arrangement of these horizons that identify a soil *series*, which is the common unit of soil mapping in Britain. A soil series consists of soils with similar profile characteristics and formed from similar geological parent materials and so in a similar climate can be expected to have the same management characteristics. A soil series is given the name of the place where it was first recognised or is particularly extensive. For example, Fladbury series identifies a poorly drained alluvial clay first recognised in the Vale of Evesham.

A *phase* may be used to subdivide a series to emphasise a local characteristic such as topsoil stoniness that is likely to have a special management implications.

In Scotland and on small scale maps in England and Wales, soil *associations* containing a number of component series, usually in a specified relationship, are the basic mapping unit.

26.5. Collecting Climatic Data

Soil interacts with climate to affect the productive capacity of a piece of land. Thus, a poorly drained heavy loam of the Brickfield series in Dyfed, receiving 1400 mm of rainfall, will be under permanent or rough pasture with a safe grazing period of only 60 days.

A similar soil in Cambridgeshire with 550 mm of rainfall will be under continuous cereal rotations.

Agro-climatic data relevant to the assessment of land quality include:

(i) *Moisture deficit* is the water requirement of a crop during the growing season that is not supplied by rainfall in that period. This deficit must be met either from reserves of moisture in the soil, or by irrigation, if yield is not to be affected.

(ii) *Accumulated temperature* is a measure of heat energy available for plant growth. It is commonly calculated for the period from January to June as the sum of the average daily temperature above 0°C on each day. Values in England range from 850°C on the north Pennines to 1600°C around Bournemouth.

(iii) *Field capacity period* is a function of rainfall and evapotranspiration, predicting the length of the period in winter when there is no moisture deficit and the soil is replete with moisture. Waterlogging will occur during this period if permeability and drainage are inadequate, and top soils are liable to compaction by stock and machinery. In parts of Dyfed the field capacity period may last from mid-September to early May, compared with early December to late March in Essex.

(iv) *Exposure* to strong winds is most detrimental along the coast and at high altitudes where it results in crop damage, particularly to trees. Forestry Commission Leaflet No. 85 explains that 'Topex' method of measuring exposure.

26.6. Soil, Climate and Productive Capacity

This section describes how soil and climatic data are brought together to assess productive capacity.

26.6.1. Droughtiness

This is an assessment of the ability of a soil to provide a growing crop with water. Soils vary in the amount of water they can store and release to a crop (known as available water). Medium and coarse sands, for example, can store relatively little water but most of what is stored is easily accessible to a growing crop. Clays, on the other hand, can store a lot of water but it is held at such high tensions in small pores that a large proportion is unavailable to the crop. Fine sands, silts and peats have the largest amounts of available water, which is partly why such soils are renowned for their agricultural quality. An estimate of the water available to crops in a particular soil in the field is made by comparison with a similar soil for which the actual quantity has been measured in a laboratory. Average amounts of available water for common soil series are given in some Soil Survey publications but precise information relating to individual profiles normally requires calculation by a soil specialist.

Because crops vary in their rooting characteristics and efficiency of water uptake, the available water value is adjusted for a particular crop. Thus, in the Hanslope series clay soil the amount of water available to sugar beet is 150 mm whereas only 110 mm is available to potatoes.

Droughtiness is calculated by subtracting the moisture deficit for a particular crop from the soil available water. A value of more than + 50 mm indicates that a soil has appreciable reserves of available water in the average season and provides some additional resistance to drought stress in dry years, and is termed non-droughty. A value of between 0 and 50 mm indicates a risk of drought stress in dry years and is termed slightly droughty. A negative value of between 0 and 50 mm, termed moderately droughty, indicates loss of yield due to drought stress in most years. A value of more than − 50 mm, termed very droughty, implies severe shortage of water in all but the wettest summers.

This calculation of droughtiness is one of the most significant interpretations of soil and climatic data for it can demonstrate how, because of water shortage, crops in some soils cannot achieve the high yields expected of them, except in a very favourable year. On some land, such as Cotswold brash, this is well known, but elsewhere the significance of drought stress may not be fully appreciated.

The droughtiness calculation also helps with the assessment of the likely benefits of irrigation by demonstrating the drought risk for unirrigated crops. Some marginal soils such as Breckland sands can become highly productive root and vegetable land once water is applied, totally transforming their economic value. However, not all soils are as well suited to irrigation as those of the Breck. Large parts of the eastern counties and Midlands have impermeable clay subsoils in which successful drainage depends upon them drying out sufficiently in summer for deep cracks to form, which will then conduct winter rainfall to mole channels and drains. Good cracks will not form if such soils are irrigated resulting in wet ground conditions at harvest time.

26.6.2. Trafficability

This is an assessment of the ease with which a soil can be traversed and cultivated and of its susceptibility to structural damage. The critical times for cultivation are in autumn, from September to November, and the spring months of March and April. The days on which land can be worked with little risk of structural damage are called 'Potential Machinery Work Days' and these represent an interaction between soil and climate.

The most useful climatic concept in assessing the trafficability of land is the field capacity period; the time during which there is no soil moisture deficit and the soil is replete with water. However, some soils are wet for longer periods than others owing to differences in texture, porosity and topography, and so the climatic data must be modified to represent more closely the likely soil conditions.

The relevant soil factors are those which relate to the ability of the soil to retain or dispose of water, as moisture content affects the bearing strength of the topsoil. These factors are the duration of wetness at various depths, the depth to any horizon which restricts water movement, and the clay content of the topsoil. They are combined to produce an adjustment to the field capacity data. In the case of a well drained, porous loamy soil there are some twenty days in autumn and ten days in the spring *within* the field capacity period when the land is trafficable. In contrast, for an impermeable clay soil there are some thirty days in autumn and ten days in spring *outside* the field capacity period when the land is not trafficable.

The effect of soil and climate on machinery work days is demonstrated by soils within the Clifton association in northern England. This soil association is dominated by seasonally waterlogged, impermeable clay loams of the Clifton series, closely associated with lighter and better drained Claverley, Salwich and Quorndon series. The association is mapped near Carlisle with 850 mm of rainfall and Darlington with only 600 mm.

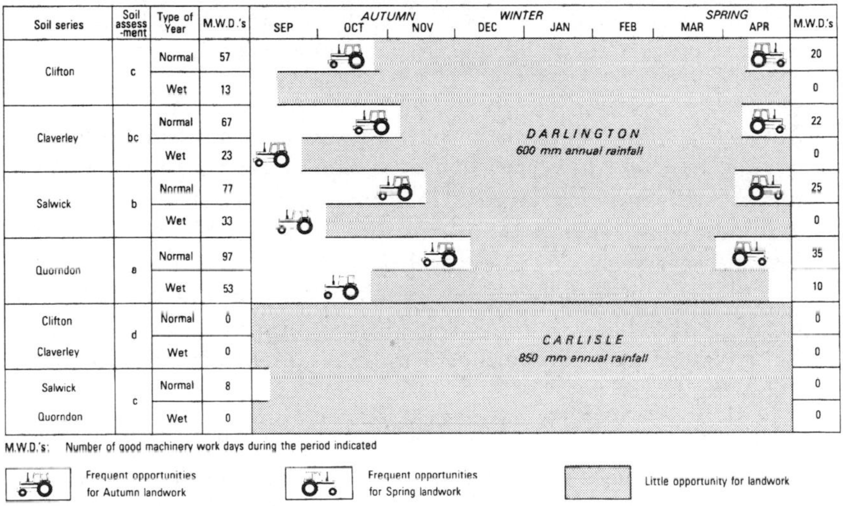

Soil series	Soil assess-ment	Type of Year	M.W.D.'s	AUTUMN / WINTER / SPRING: SEP OCT NOV DEC JAN FEB MAR APR	M.W.D.'s
Clifton	c	Normal	57		20
		Wet	13		0
Claverley	bc	Normal	67		22
		Wet	23		0
Salwick	b	Normal	77		25
		Wet	33		0
Quorndon	a	Normal	97		35
		Wet	53		10
Clifton, Claverley	d	Normal	0		0
		Wet	0		0
Salwick, Quorndon	c	Normal	8		0
		Wet	0		0

DARLINGTON
600 mm annual rainfall

CARLISLE
850 mm annual rainfall

M.W.D.'s: Number of good machinery work days during the period indicated

Frequent opportunities for Autumn landwork

Frequent opportunities for Spring landwork

Little opportunity for landwork

26.6.3. Crop Suitability

The suitability of a piece of land for any crop is a reflection of the soil and climatic factors described in the preceding sections. The number of safe machinery work days in autumn and spring are used to estimate the time available to cultivate the land and to sow and harvest the crop in the right conditions. Droughtiness classes are used to assess the ability of the soils to supply sufficient water for optimum crop growth. For certain crops specific requirements, for example sensitivity to pH, surface stoniness etc. must also be taken into account.

Four suitability classes—well, moderate, marginal and unsuited—are used and these classes are defined below as follows;

Crop suitability classes

Well suited Potential production is high and sustainable from year to year. There is adequate opportunity to establish the crop in average years at or near the optimum sowing time and harvesting its rarely restricted by poor conditions. Even in wet years (up to a frequency of about on in four) working conditions are acceptable and do not prevent crop establishment. There are sufficient soil water reserves to meet the average requirements of the crop.

Moderately suited Potential production is usually moderate or high but can be lower in years when soil water is insufficient to sustain full growth or when crop establishment is unsatisfactory owing to untimely sowing or poor soil structure. In favourable years, which may be wet years on droughty land or dry years in a generally wet climate, outstanding yields can be achieved, but production is not predictable. For root crops, harvesting can be difficult and soil structure may be damaged with consequent penalties for the following crop.

Marginally suited Potential production varies from year to year. There are considerable risks, high costs, or difficulties in maintaining continuity of output, which are due to the interaction of climate with soil properties, disease or pest problems. In some years there may be a failure to establish the intended crop. Often marginal suitability does not imply high risks in producing the particular crop but rather reflects difficulties in fitting it into a continuous farming system.

Unsuited Criteria of unsuitability vary with individual crops, but are chiefly climate, gradient and, for root crops, stoniness. Near the selected climatic limits for a crop there will be years with favourable weather which allows efficient production and other years which are too wet or cool. The climatic limits chosen, however are expected to exclude normal production of the crop at least one year in four: a higher degree of risk was judged unacceptable.

The assessments assume a moderately high level of management and appropriate underdrainage measures. Factors such as market conditions and farming systems are not taken into account.

The following are examples of the method for assessing the crop suitability of specific crops. There is a full description in each of the six Regional Bulletins of the Soil Survey and Land Research

Centre. For winter cereals the number of autumn machinery work days is particularly significant, whereas for sugar beet a combination of the number of spring (for sowing) and autumn (for harvesting) days is used.

Allocation of Suitability Classes for Winter Cereals

Machinery work days after 1 Sept.		Available water (A.P.) Moisture deficit (M.D.) in mm Over 40	>20 to 40	>0 to 20	0 to −19	−20 & lower
Over 80	Increasingly	Well	Well	Well	Moderate	Marginal
>50 to 80	restricted	Well	Well	Moderate	Marginal	
>20 to 50	workability	Moderate	Moderate	Moderate	Marginal	
20 & less	↓	Marginal	Marginal	Marginal		
		→ Increasing droughtiness				

Allocation of Suitability Classes for Sugar Beet

Machinery work days 1 Jan.–30 April plus 1 Sept.–30 Dec.		Available water (A.P.) Moisture deficit (M.D.) in mm Over 40	>20 to 40	1 to 20	1 & lower
Over 120	Increasingly	Well	Moderate	Moderate	Marginal
>90 to 120	restricted	Well	Moderate	Marginal	Marginal
>60 to 90	workability	Moderate	Marginal	Marginal	
60 and less	↓	Marginal	Marginal		
		→ Increasing doughtiness			

26.6.4. Suitability for Grassland

A similar scheme is available for grassland suitability. Grass needs sufficient water, warmth and nutrients to grow well. Soil conditions and climate regulate the supply of moisture whereas fertilizers, notably nitrogen, are applied to sustain the supply of nutrients. the suitability of land for grass depends not only on its capacity to growth the crop but also on its ability to support animals and machinery without damage to sward or soil. Interactions between climate and soil are complex and affect the potential yield, use and husbandry of grassland in different ways.

The circumstances ideal for growth are not always those that make management, grazing or cropping easy. The large potential yields that can be achieved in a moist climate may be offset by poor ground conditions. The scheme assesses potential yield, the risk of poaching and the balance between them, to indicate relative fitness for intensive grass production.

26.7. Conclusion

The systems described above, which integrate data about climate and soil, provide a systematic and scientific method of assessing the productive capacity of a holding. Much of the requisite information can be gathered or researched from existing publications, but in some situations it may be necessary to employ a consultant to collect and interpret relevant data. By whatever means the data are assembled, they should serve to replace the vague and often arbitrary assertions about land quality that have frequently prevailed in sales particulars, valuations and arbitrations. The following are some case histories where detailed soil and climatic information played a key role in a valuation. Real names and locations have not been used.

26.7.1. Sale of the Denny Estate

This 400 ha estate, owned by a pension fund, was put up for sale in 1986. The Agricultural Land Classification maps shows the land to be equally divided between Grades 3 and 4, but the agents felt this was not a fair reflection of the land quality, since most of the estate had been converted into productive arable land in recent years. The ALC grading probably reflected in part the extent of grassland on the estate at the time of survey.

The agents commissioned a detailed soil survey of the whole estate in order to accurately reassess the ALC grading. This revealed the 85% of the estate is Grade 3, only 5% is Grade 4, and the remaining 10% is Grade 2. this information probably boosted the selling value of the estate by as much as £200 per acre.

26.7.2. Rent review at Lodge Farm

A tenanted farm on an estate consisted largely of sandy soils giving lower yields of sugar beet than neighbouring farms. The agent believed that this was due to poor husbandry, possibly compaction, a claim hotly denied by the farmer who was pressing for a rent reduction. A detailed soil survey and careful examination of the soil profiles under sugar beet showed no compaction, but nevertheless the roots of some plants terminated abruptly at 35 cm depth. Soil analysis revealed than the colloidal content (a combination of clay, silt and organic matter) beneath the topsoil was only 1.04%, indicating that the rooting medium was almost pure sand; much more sandy than adjacent farms on the estate. Research in Denmark has demonstrated that sugar beet rooting is inhibited in sandy soils with a colloidal content of less that 5%. Thus, in this case the low sugar beet yields were due not to poor farming but to a poor farm with very sandy soils.

26.7.3. Dilapidations at Fenedge Farm

In 1987 the tenancy at Fenedge Farm became vacant after twelve years during which the farmer had produced carrots, potatoes and sugar beet with only an occasional cereal break. The landlord's agents were concerned that such intensive root and potato production had damaged the productive capacity of the holding through a build up of disease, compaction and depletion of certain trace elements for example copper and boron. A detailed soil survey and analysis revealed the following; moderate compaction in some fields but within plough depth and therefore easily removable; boron levels were adequate but copper was low in some fields, although no lower than is common in a district where regular applications of copper are required. The numbers of nematodes were moderate or low in all fields, as is typical of the district. However, infection with violet root rot was severe in many fields, and it was recommended that the production of carrots, potatoes and sugar beet should cease for about 5 years. Dilapidations between landlord and tenant were assessed on that basis.

Details of Soil Survey publications may be obtained from;

Publications Officer,
Soil Survey and Land Research Centre,
Silsoe Campus, Silsoe, Bedfordshire, MK45 4DT.

Soil Survey of Scotland,
Macaulay Land Use Research Institute,
Craigiebuckler, Aberdeen AB9 2QJ.

Details of British Geological Survey publications may be obtained from:

British Geological Survey,
Keyworth, Notts, NG12 5GG.

Details of MAFF publications may be obtained from;

MAFF (Publications),
Lion House,
Willowburn Estate,
Alnwick,
Northumberland, NE66 2PF.

CHAPTER TWENTY-SEVEN

Management Agreements arising from Designation of Sites of Special Scientific Interest

(see circular 4/83, 1983 by the Department of the Environment, Ministry of Agriculture, Fisheries & Food. Also Circular 6/83 Welsh Office)

27.1 Management agreements arise as a result of the designation of Sites of Special Scientific Interest (SSSI) by the Nature Conservancy Council (NCC). The sites so designated have been identified by the NCC as 'areas of land or water containing plants, animals, geological features or landforms of special interest'. The National Parks and Access to the Countryside Act 1949 Section 16 and The Countryside Act 1968 originally provided the then Nature Conservancy the right to notify planning authorities of its desire for their conservation and identification as SSSI. The 1981 Act provided the NCC the power to formally notify the owners and occupiers of such sites, the Secretary of State and local planning authorities of such designation. Most of the work of notification of such existing sites has been completed by 1987. When a site is 'notified' the owners and occupiers are told that the area is a SSSI, sent a plan of the sites boundary, a statement of the sites special interest (e.g. animals, plants geological features or landforms etc) and also a list of operations that may be damaging to the site, e.g. changes in grazing, ploughing up, etc.

Three months are given for objections (where the area concerned has not already been made a SSSI under the 1949 Act). Where objections are made consultation normally takes place between the NCC Officers and if such discussions produce agreement the objection is withdrawn. If agreement cannot be reached the matter is referred for further consideration by the NCC Council responsible for assessing SSSI cases. If the matter cannot be resolved it is then considered by a full meeting of the NCC which is comprised of members appointed by the Secretary of State for the Environment whose decision determines whether or not the notification is confirmed. Within 9 months of notifying an SSSI to the Secretary of State (the date the owner and occupier is usually notified of the proposal) the NCC must confirm or withdraw the notification. Alternatively the NCC may modify the notification.

If an owner or occupier wish to carry out any operations listed in the notification as being damaging to the site they must give the NCC 4 months notice in writing of the intention. Contravention of this requirement can lead to conviction and a substantial fine. However if consent to the operation is given or it is carried out under a management agreement the work can be carried out before the four months have expired.

27.2. FINANCIAL PROVISIONS FOR SSSI'S

The NCC may enter into management agreements, (the term of which is negotiable, although the provisions of DOE Circular 4/83 imply a 20 years or longer term agreement possibly operating in perpetuity) make grants or offer to lease or purchase the land concerned in order to enhance the conservation interest of SSSI's and also to compensate the owner and occupiers for the effect the Agreement has on their business. Compensation can take the following forms:

1. *Lump sum payments* in respect of loss of capital value because of the agreement. The amount should be equal to the difference between the restricted and unrestricted value of the owner or owner/occupiers interest calculated in accordance with the rules of assessment in respect of compulsory requisition of an interest in land as set out in Section 5 of the Land Compensation Act 1961 and in so far as there is no statutory eligibility for compensation in this respect under Section 30(2) of the 1981 Act.
2. *Annual payments* on the basis of 'net profit forgone' because of the agreement.

Note: That annual payments, only, are available to those who are occupiers and do not own the land.

These figures are negotiated by the affected persons or their valuers and the valuers to the NCC, but if not agreed, owners or occupiers can demand arbitration. For the purpose of calculating payment under a management agreement, it is to be assumed that, but for the conservation considerations, farm capital grant, (if applicable) would have been payable on the proposed agricultural operation which the claimant undertakes in the agreement not to carry on. The NCC will pay reasonable professional and legal costs for advice incurred by the owner or occupier affected on completion (net of VAT).

Conditional exemption from Inheritance Tax is available as an *alternative* to compensation payments under a management agreement. The sale of SSSI land to the NCC attracts a capital gains tax concession known as 'the doucer' if SSSI land is offered to the Government in lieu of Inheritance Tax.

Where land is subject to an agricultural tenancy, the terms of the management Agreement may have implications on the tenancy agreement, in particular the tenancy obligations relating to good husbandry. The Guidelines, therefore, suggest that the landlord should be a party to the main Agreement with the tenant, or alternatively, enter into a separate complementary Agreement. This would include an undertaking on the part of the landlord not to serve a notice to remedy contrary to the intentions of the Management Agreement. A nominal payment should normally be made to the landlord as recognition of his consent to the Agreement. However, the landlord should be entitled to continue to receive the full open market rental value of the holding as if the tenancy was not restricted from farming the land in any way.

27.4. ASSESSMENTS OF ANNUAL PAYMENTS

Individual assessment is appropriate in most cases to calculate the sums payable. Forms are completed as laid down in Annex C of Joint Circular 4.83 and 6.83. A completed version of this form is shown in the example given. Occasionally in large SSSI's where many owners are involved a standard payments system might be suggested to facilitate negotiation and form an apparent equitable basis of compensation.

27.5. FORESTRY OPERATIONS

Conditions which may be sought in forestry operations are likely to cover:

(a) *Outright Prohibition* of planting of bare land and felling of woodland. This will be for a stated number of years.
(b) *Modification of Management Practice* covering new planting and re-stocking, inc. choice of species, maintenance operations and clear felling.

Compensation here will alternatively be either

1. *Lump Sums* determined by assessment of revenue forgone, based on a comparison of discounted streams of expenditure and income and the period of the management agreement and calculated,
   (a) with, and
   (b) without
   the constraints imposed by the management agreement. The rate of discount used is to be used in calculating present worth and agreed between the representative body of private foresters and the NCC.
   or
2. Payment based on the depreciation in value of the land or woodlands concerned, calculated having regard to the rules of assessment as set out in Section 5 of the Land Compensation Act 1961.

Where *annual payments* are required these should be derived from the lump sum (1 above) rentalised or amortised to produce a flow of annual payments based on an estimate of current market rates of interest over the terms of the management agreement.

The depreciation in value of land method is generally preferred on account of the complexity of the calculation in the net discounted revenue basis and the very different results that can be obtained as a result of variations in the discount rate selected.

An even more satisfactory approach as an alternative to compensation type payments may be payments towards woodland management including regeneration, planting and fencing and which will take into account nature conservation requirements.

27.6. RECREATIONAL MATTERS

The Financial Guidelines for Management Agreements relate primarily to agriculture and forestry claims.

There are sometimes claims for loss of recreational activities which must be processed on similar lines to the operation of the Guidelines but outside their statutory provisions.

Claims for loss of sporting interests do not occur widely as often there is compatibility between shooting, fishing and nature conservation. Problems may only arise indirectly due to damage to ground cover by intensive game rearing or intensive stocking of fishing areas.

The more usual claims are for other activities such as motor cycle trials, motor rallying or war games. These activities are controlled to a considerable extent by planning restrictions. However, the provisions of the General Development Order 1988 entitle such activities to be carried out without a planning application for up to 14 days in any calendar year.

These requirements are specified in Schedule 2, Part 4, Class B2 of the GDO which applies a 14 day rule to:

(i) The holding of a market
(ii) Motor car and motor cycle racing including trials of speed and practising for these activities.
(iii) Clay pigeon shooting.

N.B. The rules relating to clay pigeon shooting have been further revised during 1989—a 28 day period is now operational.

It is possible that (where such activities on a 14 or 28 day basis will be harmful to nature conservation requirements), an Article 4 Direction may be imposed by the Local Planning Authority or DoE. This direction effectively will bring the activity within planning control and require a planning application (i.e. the GDO relaxations will be withdrawn). Payments may be made by the Local Planning Authority in respect of depreciation in value of land if planning permission is refused. This will be assessed in relation to the open market capital value of the land rather than the annual profits.

27.7. EFFECTIVE DATE

Where a notice to carry out an operation is made, the commencement date or, 'effective date' of the Agreement shall be 4 months after receipt of the notice. The valuation occurs after the effective date. Where completion of the Agreement occurs after the effective date interest may be paid at the rate prescribed by the regulations under S32 of the Land Compensation Act 1961 (although interest payments may be with-held if there has been delay in the submission of a claim).

27.8. HEADS OF TERMS OF MANAGEMENT AGREEMENTS

The terms of years of such Agreements is negotiable.

Where a term of 3 years or less is proposed, and payment is less than £5,000 per annum the Agreement may be completed without a sealed document but will still normally be registered as a land charge. Longer term Agreements are usually completed under seal.

The terms of Agreements are very widely empowered under the Act.

In essence Agreements can be:

(a) Compensatory in nature.
(b) Positive nature conservation Agreements whereby management works such as scrub control are required.
(c) Payments 'in kind' situation where e.g. NCC may agree to erect a fence in lieu of a compensation type payment to an owner (but payments made should not be in excess of the payments which would have been due if any Agreement of a compensatory type had been envisaged).

An owner or occupier might wish to make their own contribution by accepting less than the full payment.

There is often a gain to an owner and NCC in entering a 'positive management' type Agreement.

An owner can agree to fencing work, water supply or scrub control being carried out by NCC rather than take annual payments for restricted used of the land. Annual payments are of course subject to income tax and must cease if a conditional inheritance tax exemption undertaking is entered into.

The Agreement usually is completed on a standard format and is in 3 parts:

1. The Agreement (under seal if for over 3 years or over £5,000 per annum).
2. The Agreed Management Policy.
3. The Management Plan.

The Agreement contains legally binding covenants relating to such items as:

1. Names of parties.
2. Description of land.
3. Payments and reviews.
4. Term of years.
5. Termination and repayment provisions in respect of cessation of SSSI designation, breach or conditional exemption from Inheritance Tax.
6. Arbitration provision in the event of dispute.
7. Access.

Normally access would be required only for NCC members and staff and other persons being scientists engaged in genuine scientific study or research or persons with a special interest in nature conservation.

The Agreement Management Policy contains provisions such as:

1. Continuing liaison between the parties. Matters to be discussed such as scientific work, erection of fences etc.
2. Lists the potentially damaging operations and requires that these are not carried out without the prior agreement of NCC. (This is a result of the requirements of S28 (6)(b) of the 1981 Act.)
3. Shooting provisions (including list of species which can be controlled, both game and pests.)

The Management Plan which usually has to be tailor made to fit the site will set out objectives as:

1. Grazing and mowing dates, type of stock and stocking rates.
2. Woodland management (planning, thinning, felling).
3. Drainage limitations.
4. Positive management such as bracken or gorse control.

Normally NCC will not be able to make any payments until time as an Agreement is completed. In case of long Agreements some delay is inevitable whilst legal formalities are being completed. In such cases NCC may offer a Short Term Agreement so that 'interim type' payment may be made to an owner of occupier.

These Agreements are usually fairly simple and basic documents which are usually for a term of 9 months.

Although the term of years is negotiable, NCC would normally expect an owner to sign and Agreement of 20 years duration where capital works such as fencing are being financed by NCC in the expectation that such works if carried out satisfactorily will be serviceable on such a time scale.

27.9. EXAMPLE CLAIM

ANNUAL PAYMENTS FOR NET PROFITS FORGONE.

1. If the owner or occupier chooses individual assessment, then he should provide the offeror, if possible within one month or being requested with the required information as set out in the following forms. In addition, he should supply:

— A plan showing the total area farmed by him, including the area proposed for improvement;
— If requested by the offeror, confirmation of his financial ability to carry out the proposed operation: e.g. a bank manager's letter;
— Any other relevant information that the offeror may request.

2. The offeror will either accept the sum claimed or enter into negotiations with the offeree.

3. The information given in the forms should be as complete as possible because it will form the basis for negotiations. Nevertheless, other factors may emerge which will need to be included.

IN CONFIDENCE

Information submitted in support of a claim for annual payments based on an individual assessment of net profits forgone.

I claim the sum of £..875........... as detailed in Part 5 of this form

Signed....J..P..SMITH.. Date..8th..January...1990

Information on the claimant

Full name..JOHN..PERCIVAL..SMITH...

Address:...WYND..END..FARM,..BROMARSH,..HEREFORD................ Postcode.HR9..6ER.........

If the address of your farm is different, please enter it here ....

.... Postcode.......................

Please give the name(s) and address(es) of your professional adviser(s)....................................

....MESSRS..HANDALL..&..FIRM,..CHARTERED..SURVEYORS,..THE..WALKS,..LEDBURY,.............

....HEREFORDSHIRE... Postcode(s) HR6..4NB........

Is your interest in the land freehold* ☑ or leasehold ☐ ? *(Please tick the appropriate box).*

If you lease the land, state the name and address of your landlord, the terms of tenancy, rent passing and date of last review; and confirm that the landlord has been notified as required by paragraph 7 of the guidance..

....

....

....

Please detail any restrictions, easements† or rights of way affecting the land

.NIL...

Please detail any land charges or mineral rights ....

.NONE..OF..WHICH..I..AM..AWARE..

In Scotland:* = dominium utile. † = servitudes.

County (in Scotland, Region) HEREFORD..&..WORCESTER..........

1—Information on current crops, livestock and labour

Please complete these summaries of the cropping area, number of livestock and labour engaged for the whole farm.

Crops, etc	Hectares*
Wheat	20
Barley	20
Oats	10
Other cash crops (please specify)	
Permanent grass	40
Temporary grass	20
Fodder crops	5
Total crops and grass	115
Rough grazing	4
Buildings, Roads, Woodlands etc.	2
Total farm area	121

Stock	Number
Dairy cows	
Beef cows	30
Beef heifers	100
Fattening/rearing cattle	200
Ewes	
Sows	
Fattening pigs (over 20kg)	
Laying hens	
Broilers	
Other stock (please specify)	

Labour	Number
Farmers and partners	2
Farmers' and partners' spouses doing farmwork	
Regular full-time workers	1
Regular part-time workers	
Seasonal and casual workers	

**1 acre = 0.40 hectares*

2—Description of the current situation and the proposed improvement

What is the extent of the area you propose to improve? | 10 acres | 4 hectares*

1 acre = 0.40 hectares

Please describe:

—The current situation (*eg "The area is mainly rough grazing (mostly heather), supporting about 50 ewes and lambs throughout the summer. Lambs are sold as stores"*)

Rough grazing land with wet hollows and patches of rushes

—The proposed improvement (*eg "The area is to be drained, limed, fertilised and reseeded"*)

Drain plough and reseed to use for grazing and silage

—The proposed practice after improvement (*eg "the stocking rate would be raised to 200 ewes during the summer. Lambs would still be sold as stores"*)

The land would produce 10 tons silage per acre after improvement (one cut)
and produce improved aftermath grazing

3—Annual financial effect of the proposed improvement.

Include all revenue and cost changes (variable and fixed) which would be expected as a result of the improvement. Use current values and costs.

If there are substantial year by year differences (eg because of phasing of improvements) use a fresh sheet for each year. The final sheet should reflect the full effects.

Budget for 19.90.............

Extra Variable and Operating Costs

Number and type of Units *eg 2 tonnes compound fertilizer*	Cost/Unit £	Total Cost £
3 cwts/acre fertiliser (compound 20:10:10)	120/ton	180
Silage cutting, loading clamping	30/acre	300
	Total	480 (a)

Extra Revenue

Number and type of Units *eg 10 heifers*	Value/Unit £	Total Revenue £
10 tons silage/acre	22 ton	2200
10 weeks aftermath grazing	20/acre	200
	Total	2400(c)

Revenue Forgone

Number and type of Units *eg 10 lambs*	Value/Unit £	Total Revenue £
10 head of cattle (summer grazing)	50	500
	Total	500 (b)

Variable and Operating Costs Saved

Number and type of Units *eg 200 kg N*	Value/Unit £	Total Cost £
Topping of rushes spraying of thistles	14/acre ($2\frac{1}{2}$ acres)	35
	Total	35 (d)

Annual Benefit from Proposed Improvements (before fixed cost adjustment)

(c + d) − (a + b) = £ 1455 (A)

(2400 + 35) - (480 + 500)

4—Capital requirements of the improvement *(if any)*

Please give details of any additional capital expenditure for buildings, drainage, fencing, machinery etc.

Items			Total Cost £		*Annual Charge Factor	Annual Charge Net of Grant £
No	Description	Gross unit Cost (£)	Gross	Net of Grant		
	10 acres land drainge	400	4000	2800	-163	456
	10 acres reseeding	60	600	600	-206	124
		Total				580 (B)

5—Amount claimed

Estimated Net Annual Profits Forgone — A − B = £ 875 (C)

(1455 - 580) = 875

*The multiplying factor used should reflect the economic life of the item and the appropriate interest rate: eg the factor is 0.163 for an item with a 10 year life at 10% interest.

CHAPTER TWENTY-EIGHT

Glossary of General Information

28.1 STORAGE REQUIREMENTS (Approximate only)

Bulk grain in store @ 16% moisture:

	m^3/tonne	*Kg/m^3*	*Cu ft/tonne*
Wheat	1.3	770	46
Barley	1.45	690	49
Oats	1.96	510	67
Beans	1.25	800	42
Oil seed rape	1.4	714	50
Maize	1.3	770	46
Roots (in bulk)			
Mangolds	1.78	562	64
Swedes	1.78	562	64
Turnips	1.9	510	67
Potatoes	1.55	645	55
Fodder beet	1.78	562	64
Other commodities			
Hay			
Small bales	8–10	100–125	282–353
Wheat straw			
Small bales	11–12	83–90	388–423
Barley straw			
Small bales	13	77	459
Silage (approx only)			
Wilted grass	1.25	800	44
Tower silage/haylage (45–50% DM)	1.55–1.77	560–640	54–62
Maize silage	1.3	770	46
Concentrates			
Meal	2	500	71
Cubes/Nuts	1.6	625	56

28.2 SEEDING RATES AND APPROX COSTS 1990

Cereal seed—commercial/dressed	*Per ha*		*Per acre*	
Winter wheat	185 kg	£37.00	75 kg	£15.00
Winter barley	185 kg	£37.00	75 kg	£15.00
Winter oats	185 kg	£44.50	75 kg	£18.00
Spring wheat	220 kg	£48.50	90 kg	£20.00
Spring barley	185 kg	£48.00	75 kg	£19.50
Spring oats	210 kg	£58.75	85 kg	£23.80
Potatoes (early) (once grown)	3,700 kg	£925.00	1,500 kg	£375.00
Potatoes (main crop) (once grown)	2,500 kg	£425.00	1,000 kg	£170.00
Swedes (graded seed)	$1\frac{1}{2}$ kg	£24.70	$\frac{1}{2}$ kg	£10.00
Mangolds (graded seed)	$6\frac{1}{2}$ kg	£39.00	$2\frac{1}{2}$ kg	£15.00
Turnips (graded seed)	$1\frac{1}{4}$ kg	£5.00	$\frac{1}{2}$ kg	£2.00
Fodder beet (pelleted seed)		£76.50	(55,000 seeds)	£31.00
Stubble turnips	$5–7\frac{1}{2}$ kg	£15.00	2–3 kg	£6.25
Rape (broadcast)	$7\frac{1}{2}–12\frac{1}{3}$ kg	£15.00	3–5 kg	£6.00
Kale—marrow stem (graded)	$1\frac{1}{4}$ kg	£6.00	$\frac{1}{2}$ kg	£2.50
Kale—hybrid (graded)	$1\frac{1}{4}$ kg	£10.00	$\frac{1}{2}$ kg	£4.00
Cabbages (graded)	$1\frac{1}{4}$ kg	£100.00	$\frac{1}{2}$ kg	£40.00
Maize		£86–£123.00	(45,000 seeds)	£35–£50
Grass seeds (long leys)	33–37 kg	£56.00	13–15 kg	£23.00
Forage peas	86 kg	£41.25	30 kg	£14.40
Lucerne	25 kg	£75.00	10 kg	£30.00
Oilseed rape	$6–8\frac{1}{2}$ kg	£28.15	$2\frac{1}{2}–3\frac{1}{2}$ kg	£11.40
Field beans	250 kg	£95.00	100 kg	£38.00
Field peas	250 kg	£90.00	100 kg	£36.00
Linseed		£89.00		£36.00

28.3 CROP YIELDS (AVERAGE FOR HEREFORDSHIRE)

	Per ha	*Per acre*
Winter wheat	$7\frac{1}{2}$ tonnes	3 tonnes
Spring wheat	5 tonnes	2 tonnes
Winter barley	$6\frac{1}{4}$ tonnes	$2\frac{1}{2}$ tonnes
Spring barley	$4\frac{1}{3}$ tonnes	$1\frac{3}{4}$ tonnes
Winter oats	$7\frac{1}{2}$ tonnes	3 tonnes
Spring oats	5 tonnes	2 tonnes
Hay (first cut)	5–6 tonnes	$2–2\frac{1}{2}$ tonnes
Grass silage (first cut)	14–20 tonnes	6–8 tonnes
Potatoes (main crop—not irrigated)	40 tonnes	16 tonnes
Swedes (precision drilled)	50–75 tonnes	20–30 tonnes
Mangolds	75–100 tonnes	30–40 tonnes
Sugar beet (washed)	40 tonnes	16–17 tonnes
Cabbages	75–100 tonnes	30–40 tonnes
Fodder beet	85 tonnes	35 tonnes

28.2 FERTILISERS

Fertilisers are measured in units of N (Nitrogen), P_2O_5 (Phosphate), and K_2O (Potash (N.P.K.), 1 unit is the percentage of one 50 kg bag of fertiliser i.e. a compound with an analysis of 20:10:10 = 20 units of Nitrogen, 10 units of Phosphate and 10 units of Potash. Thus, if a farmer is applying 5 × 50 kg of 20:10:10 per ha (2 × 50 kg per acre) he is applying:

100 units Nitrogen per ha (40 units of N per acre)
50 units Phosphate per ha (20 units of P_2O_5 per acre)
50 units Potash per ha (20 units of K_2O per acre)

Phosphate, for the purpose of calculating unexhausted manurial values, is further divided into soluble and insoluble, and this is invariably shown on the bag.

Some intensive farmers use a great deal of Nitrogen and there are cases of as much as 400 units of Nitrogen per acre (1,000 units per ha) being used by dairy farmers, at average intervals of three weeks during the growing season. Similarly, some intensive and specialist cereal growers apply three or possibly four top dressings of Nitrogen at three-weekly intervals in the spring—often 170 units and more. Nitrogen is rarely applied in the winter as rainwater leaches (washes) it out of the soil. It is always wise to have the soil tested for Phosphate and Potash levels. Basic slag, when obtainable, is very good for grassland and is generally applied at the rate of ½ tonne per acre (1.25 tonnes per ha). It encourages clovers, which in turn improves the fertility by naturally fixing Nitrogen.

Straight fertilisers are straight Nitrogen (e.g. Nitram), Phosphate (e.g. Superphosphate), Potash (Muriate or Sulphate of Potash).

Compound fertilisers are compounds or combinations of these, e.g. 20:10:10. as described above.

Typical fertiliser applications are:

Winter corn 5–6 × 50 kg low Nitrogen compounds (e.g. 9:25:25) per ha (2–2½ × 50 kg per acre).
Spring dressing of, say, 5 × 50 kg Nitrogen (e.g. Nitram—34.5% N) per ha. (2 × 50 kg per acre). Sometimes, further Nitrogen will be applied 3–4 weeks later (and possibly more) up to a total of 420 units per ha (170 units per acre).

Spring Corn 5–6 × 50 kg compound (e.g. 20:10:10) per ha (2–$2\frac{1}{2}$ × 50 kg per acre), together with further top dressing of Nitrogen, as above.

Potatoes $1\frac{1}{4}$ tonnes per ha ($\frac{1}{2}$ tonne per acre) high Potash, low Phosphate compound e.g. 20:8:14 before planting.

Sugar beet Salt (e.g. Beetrox) in the autumn/winter at $\frac{1}{2}$ tonne ha (4 × 50 kg per acre). Before planting, 0.75 tonne compound (manufacturers have their specialist fertilisers) per ha.

Grassland 5 × 50 kg grassland compound e.g. 20:8:14 per ha (2 × 50 kg per acre) with 3–4 further applications of Nitrogen (say 50 units a time) in the season, depending on the intensity of the enterprise.

28.5 LIME

This is not a fertiliser, but is an essential nutrient and is employed to counteract soil acidity. It is necessary for nitrification, nitrogen fixation etc. The intensity of soil acidity is measured in terms of pH units. The optimum pH for most crops is 6.5. Below this, lime requirements can vary considerably according to soil type. The heavier the soil the more lime is required many severely deficient cases requiring an application of $7\frac{1}{2}$ tonnes per ha (3 tonnes per acre) upwards. However, it is not practicable to apply more than $7\frac{1}{2}$ tonnes per ha (3 tonnes per acre) per application. The cost in 1990 is variable, but, dependent on the amount supplied (the more supplied the cheaper it will be) and haulage costs, the average in Herefordshire is from £12 per tonne to £13 per tonne.

28.6 SPRAYS

These are of the four types generally used in agriculture and the current (1990), approximate only, cost, applied.

(1) *Herbicides*—for weed control viz:
  (a) Pre emergence e.g.
    Sencorex in potatoes (£38 per ha)
    Prebane in cereals

Goltix in sugar beet £35–45 per ha) (low dose programme—usually more than one application)
Remtal in peas £26 per ha)
Commodore in oil seed rape (£44 per ha)

(b) Non selective contact e.g.
Gramoxone
Cleansweep
Scythe

Non selective translocated i.e. passes to and kills the root system e.g.
Round-up

(c) Selective contact and/or translocated e.g.
MCPA or CMPP in cereals
Asulox in pasture
Betanal E in sugar beet (£35–45 per ha) (low dose programme—usually more than one application)
Kerb in oil seed rape (£45 per ha)

(2) *Fungicides*—for controlling fungal diseases e.g. mildew, eyespot, rust in cereals and blight in potatoes.
Preventative or curative sprays *in cereals* e.g.
Dorin £24 per ha.
Tilt (£24 per ha)
Sportak (£26 per ha)
Calixin (£17 per ha)

in potatoes e.g.
Dithane (£12.50 per ha)
Fubol (£23 per ha)
Patafol (£20 per ha)

Alternatively, seed dressing e.g. in cereals
Ferrax for broad spectrum—disease control in barley (@ 190 kg seed/ha = £16.90 chemical cost only)
Baytan for broad spectrum diseases in wheat and barley (@ 190 kg seed/ha = £13.70 chemical cost only)

(3) *Insecticides*—for controlling aphids and other insects and prevention of virus diseases in cereals and potatoes e.g.
Metasystox (£8.50 per ha)
Aphox (£15.00 per ha)

(4) *Growth regulators*—applied to cereals to shorten and stiffen the straw to avoid lodging e.g.
Cycocel (£10.80 per ha)

Terpal (£26.00 per ha)
Cerone (£23–£27.50 per ha)

The only reduction in feeding value is in reduction of the bulk produced. There is nothing proven that these products reduce actual feeding value.

28.7 EXAMPLE OF 1990 COSTS OF MAKING GOOD DILAPIDATIONS

These are intended as a guide only and are costings current in Herefordshire and Gloucestershire in 1990. V.A.T. is also payable where applicable.

Note: These figures should be used with care having regard to all the circumstances e.g. some ditches are easier to clean out than others and some hedges easier to lay.

28.7.1 Hedging and Fencing

(a) Cutting and laying very high hedges including supplying stakes and clearing up afterwards. £2.30–£3 per yard.

(b) Cutting and laying medium and small hedges supplying stakes clearing up afterwards. £2–2.30 per yard.

(c) Providing and planting two rows of quickthorns and protecting on both sides with pig netting and single stand barbed wire £8 per m.

(d) Trimming hedges with flail type trimmer. The cost is from £10 to £14.50 per hour.
Approximately 300 m of small hedge can be done, both sides, in an hour—3½–4 p per m. Some hedges are more difficult to gain access to and are high and might need cutting twice. In these cases, it is difficult to do more than 200 m per hour—5½–6 p per m.
}
This is where all the hedges on the farm are done at one time. If a claim is in respect of an isolated hedge cost are doubled or trebled.

(e) Siding up high hedges with flail trimmer—say 500 m per hour—2½–3 p per m.

28.7.2 Fencing and Gates

All erected to the Ministry of Agriculture standards and all timber tanalised and treated.

(a) Pig netting (medium) on tanalised posts at 2.4 m centres with two strands of barbed wire £1.90–£2.10 per yard, £2.07–£2.28 per m. (at least 1.2 m high).
(b) Post and 4 rail fencing at least 1.2 m high. Sawn posts and rails (tanalised) £6 per yard, £6.54 per m.
(c) Post and 4 rail fencing but posts round and rails half round £5 per yard, £5.45 per m.
(d) Barbed wire fence, 2 strands, with treated softwood posts at 3 m centres £1.30 per yard, £1.41 per m.
(e) Barbed wire fence, single strand, 5 m centres on treated softwood posts 80p per yard, 87p per m.
(f) 3.6 m galvanised medium gauge metal gate complete with hanging and shutting and all fittings £120–£130 each.
(g) 3.6 m oak gate with posts and all fittings £200 each.
(h) 4.5 m medium gauge galvanised metal gate and posts and all fittings £130–£150 each.

28.7.3 Ditching and Drainage

(a) *Ditches*
Cleaning out a small ditch (by hand) 80p–100p per yard, 87p–109p per m.
Mechanically cleaning out silted and filled up deep ditches and depositing spoil at sides.
$1\frac{1}{2}$ m wide at top $\frac{2}{3}$ m at bottom—£1.20–£1.40 per yard, £1.30–£1.52 per m.
$2\frac{1}{2}$–3 m wide at top—£1.60–£2 per yard, £1.74–£2.18 per m.
(b) *Land Drainage*
Excavate trenches, supply and lay and backfill
160 mm plastic pipe £2.05 per m run
150 mm plastic pipe £1.90 per m run
100 mm plastic pipe £1.20 per m run
80 mm plastic pipe 90p per m run
60 mm plastic pipe 85p per m run
Junction pipes £2.50 each

Backfill for 150 mm pipe 90p per m run
60–100 mm pipe 70p per m run
Outfalls £20 each
Delivery of plant £60

(c) *Dragline*
Large machine £20 per hour
Smaller machine (JCB) £12–£15 per hour

(d) *Mole Drainage*
£50–£55 per ha (£20–£22 per acre)

28.7.4 **Material Costs** (all plus V.A.T.)

Sheep netting £35 per roll (50 m)
Pig netting (medium gauge) £21.50 per roll (50 m)
Barbed wire (two wires) £15 per roll (200 m)
Plain galvanised wire—high tensile £22 per roll (410 m)
Straining posts 2.1 m long tanalised
120–150 mm dia.—£6.50 each
Intermediate posts 1.7 m long tanalised
75 mm dia.—90p–£1.20 each.
100 mm dia.—£1.30p each.
Chestnut fencing stakes 85p–90p each.

Rails:
3.7 m long tanalised
Half round (100 mm) £2.50–£3 each.
Rectangular 75 mm × 36 mm £2.40–£2.50 each.

Gates:
3.6 m medium gauge primed £36–£40 per unit.
3.6 m medium gauge galvanised £38–£42 per unit.
4.5 m heavy gauge primed £55–£60 per unit.
4.5 m medium gauge primed £45–£50 per unit.
4.5 medium gauge galvanised £55–£60 per unit.
Labour for erecting gates (concreting not included) £40–£50 per unit.
Ironwork for gates £15 per unit.

Gate posts:
Iron gate posts (primed) £30–£32 per unit.
Ex railway sleeper posts £10 per unit.

28.7.5 Basic Costs

Worker day £39.80 per day
Two wheel drive tractor (say 75 ph and not including implements) £40 per day

28.8 TIME LIMITS (under 1986 Act unless otherwise stated)

(1) *Claim for 1986 compensation provisions to apply to 'Tenant Right' Claim where tenancy commenced before 1st March, 1948—Schedule 12, paras 6–9.*

Tenant must elect, in writing, before tenancy terminates that compensation provisions specified in Part II of Schedule 8 of the 1986 Act shall apply to him.

(2) *Termination of Tenancy Claims*—Landlords and Tenants Claims—Section 83(2). Notice of intention to claim must be served before 2 months have expired from the termination date.

Eight months are allowed from the termination date to settle the claims. If not settled within this period, the matter is then referred to arbitration (Sections 83(4) & (5).

(3) *Variation of Rent*—Section 12.

Unless agreed otherwise in the tenancy agreement, or lease, rental cannot be varied earlier than the expiry of three years from the following dates:

(i) date tenancy commenced
(ii) date of previous alteration or date an arbitrator directed that the rental should not be changed.

A notice requiring arbitration on the rental must be served in the case of a yearly tenancy, at least one year before the anniversary date of the tenancy.

(4) *Tenants Fixtures*—Section 10

These can be removed at any time during the continuance of the tenancy or before the expiration of two months from the termination of the tenancy provided the tenant has paid all rent and satisfied his other obligations to the landlord and at least one month before the exercise of the right and termination of the tenancy given the landlord written notice of his intentions to remove the fixture. The landlord can, before the expiration of the notice, give a written counter-notice electing to purchase the fixture or building at the fair value thereof to an incoming tenant.

(5) *Disturbance*—Section 60
If claim is to be made for more than one years' rent, tenant must give landlord an opportunity of valuing the stock and must give written notice of intention to make such a claim not less than one month before termination of tenancy followed by a claim under S.83 within 2 months of the date of termination.

(6) *Deterioration*—Section 72
Notice of intention to claim for *deterioration* (as distinct from dilapidations) must be served not later than one month before the tenancy terminates.

(7) *Arbitration*—Schedule 11
Statements of case must be lodged with the arbitrator within 35 days of his appointment.

An arbitrator has 56 days from his appointment to issue his award unless this period has been extended on application to the Minister of Agriculture.

(8) *Claim for Tenancy Succession on Death of Tenant*—Agriculture (Miscellaneous Provisions) Act 1976 Part II
Application for tenancy must be lodged with the Agricultural Land Tribunal within three months beginning with the day after the date of death of the tenant.

(9) *Notice to Remedy Breaches*—Agricultural Holdings (Arbitration on Notices) order 1987 (S.I. 1987/710)
If a tenant wishes to contest liability to do any work specified to be done in a Notice to Remedy, he must serve a notice in writing on his landlord requiring arbitration within one month after the service on him of the Notice to Remedy.

(10) *Determination of standard milk quota and tenants fraction before end of tenancy*. Agriculture Act 1986 Part III Section 10. Landlord or tenant may at any time before the termination of the tenancy serve a written notice on the other demanding that the determination of the standard quota for the land and the tenants fraction shall be referred to arbitration (under the provisions of Sect. 84 of the 1986 Act).

28.9 METRIC CONVERSION FACTORS

METRIC TO IMPERIAL

Weight

1 kilogram = 2.205 lbs
1,000 kilograms = 0.984 ton
1 tonne = 1,000 kg (2,204 lbs)

Length

1 millimetre = 0.04 inch
1 centimetre = 0.39 inch
1 metre = 3.279 feet
1 metre = 1.094 yards
1 kilometre = 0.6214 mile

Volume

1 litre = 0.22 gallon
1 litre = 1.76 pints
1 cubic metre = 1.307 cubic yards
1 cubic metre = 35.315 cubic feet

Area

1 hectare (10,000 sq. m.) = 2.471 acres
1 sq. metre = 1.196 sq. yards
1 sq. metre = 10.764 sq. feet

IMPERIAL TO METRIC

Weight

1 lb = 0.454 kilogram
1 score 9.07 kilogram
1 cwt = 50.80 kilograms
1 ton (2,240 lbs) = 1,016 kilograms

Length

1 inch = 25.4 millimetres
1 inch = 2.54 centimetres
1 foot = 0.305 metre
1 yard = 0.914 metre

Volume

1 gallon = 4.546 litres
1 pint = 0.568 litre
1 cubic yard = 0.765 cubic metre
1 cubic foot = 0.028 cubic metre

Area

1 acre (4,840 sq. yards) = 0.405 hectare
1 sq. yard = 0.836 sq. metre
1 sq. foot = 0.093 sq. metre

Index

A

B

C

D

E

F

G

M

N

O

P

Q

R

S